DATE DUE

JA 8 '93		
DE 9'99		
MY 23 06		

DEMCO 38-296

Origins of Nature's Beauty

Number Nineteen
Corrie Herring Hooks Series

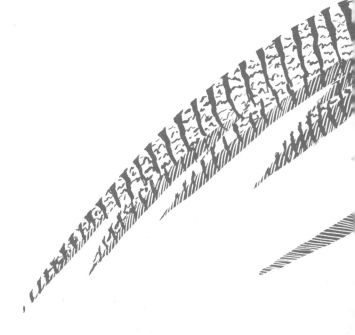

Essays by Alexander F. Skutch

Origins of Nature's Beauty

Illustrations by
Dana Gardner

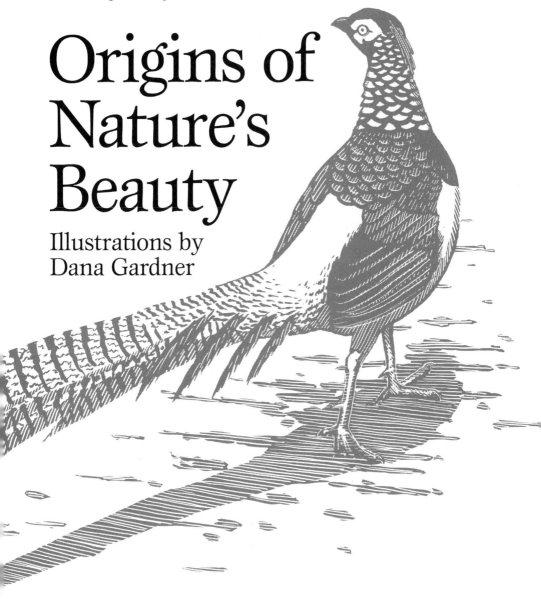

University of Texas Press   Austin

First edition, 1992

Requests for permission to reproduce material from this work should be sent to
Permissions, University of Texas Press, Box 7819, Austin, TX 78713-7819.

⊗ The paper used in this publication meets the minimum requirements of American
National Standard for Information Sciences — Permanence of Paper for Printed Library
Materials, ANSI Z39.48-1984.

Library of Congress Cataloging-in-Publication Data

Skutch, Alexander Frank, 1904–
    Origins of nature's beauty / essays by Alexander F. Skutch ;
illustrated by Dana Gardner. — 1st ed.
        p.   cm. — (Corrie Herring Hooks series ; no. 19)
    Includes bibliographical references (p.   ) and index.
    ISBN 0-292-76037-X
    1. Birds — Behavior.   2. Sexual selection in animals.   3. Nature
(Aesthetics)   4. Courtship of animals.   5. Birds — Evolution.
    I. Title.   II. Series.
    QL698.3.S557   1992
    598.256'2 — dc20   91-884
                                                                          CIP

Publication of the color plates has been assisted by a grant from the Bleitz Wildlife
Foundation.

To
John William Hardy,
whose records have brought
the beauty of bird song
to many people

Contents

Illustrations

Black and White

Color

Tables

Acknowledgments

I began to look into the literature on sexual selection in the Peabody Library of Baltimore and the Museum of Comparative Zoology of Harvard University while holding a Fellowship of the John Simon Guggenheim Memorial Foundation in 1952. The Western Foundation of Vertebrate Zoology, Dana Gardner, and Allen M. Young of the Milwaukee Public Museum sent me photocopies of papers unavailable to me in the valley of El General. Bruce Beehler gave helpful suggestions for improving the manuscript, especially the chapter on birds of paradise, which he knows in Papua New Guinea and I have never seen. The Western Foundation of Vertebrate Zoology and the American Museum of Natural History kindly loaned the artist specimens of birds for making his drawings. Gustavo Morejon loaned specimens of insects. The painting of Temminck's Tragopan was based on a photograph by David Rimlinger. A generous contribution by the Bleitz Wildlife Foundation has enabled us to adorn this book with color plates. To all these institutions and individuals, the author and the artist are grateful.

Introduction

The concept of evolution pervades modern thought and enters prominently into discussions of major biological problems. Despite the countless books and articles written about evolution, certain of its aspects remain puzzling. That evolution has produced much that is admirable as well as much that is horrible in the living world scarcely anyone will deny. The horrors are readily explicable by widely accepted evolutionary principles; developments that most deserve our approbation are often more perplexing. How can a process dominated by quantity yield quality that is more than adaptation to an environment or a life style? How can aesthetic and moral values emerge from a process indifferent to these high values? More than a century ago, Darwin undertook, with his usual perspicacity, to answer these questions. Since his day, so many pertinent facts have been discovered by field naturalists in many parts of the world, his views have been so variously criticized, that the time seems ripe for a new comprehensive survey of the subject, in the light of the most recent research.

Contemporary biology, saturated with evolutionary theory, measures the fitness of organisms simply by their capacity to produce viable progeny. This, of course, includes their ability to survive in the struggle for existence, for unless they survive to maturity they cannot reproduce. Organisms of all kinds, especially animals, appear to be engaged in unremitting competition to leave more descendants than other individuals of their species, for the most prolific lineage will eventually, by the weight of numbers alone, supplant the less prolific. In this ceaseless rivalry to multiply progeny, nothing else appears to matter, neither the wholeness of the organism nor its relations with surrounding creatures, neither longevity nor a satisfying life. To multiply one's own genes, so that they may replace competing genes, is, according to contemporary theory, the single evolutionary imperative.

No matter by what sacrifice of its own qualities or what outrages to other creatures an organism augments its fitness, its lineage will increase through the generations. This relentless impulsion to survive and multiply at any price to self or others has produced a vast array of internal parasites, blind, limbless organisms that batten on the tissues and body fluids of their hosts, and of external parasites that torture the animals whose blood they suck. It has reduced a reptilian stock to limbless, voiceless snakes, stripped to the bare essentials of a relentlessly predatory existence. It has equipped animals with fangs, horns, talons, and poison glands for overpowering prey whose flesh they tear or, in the case of males, fighting for access to females needed for the propagation of their genes. It has developed in certain mammals, including lions and some primates, the habit of mercilessly destroying the sucklings of females that they wrest from other males, to the end that the mother of the slaughtered young may the sooner become pregnant with the murderer's progeny. It is responsible for the harsh aggressiveness, the violent, disturbing passions that infect the higher animals, none more distressingly than humans. In short, the evolutionary impulsion to increase fecundity, technically known as fitness, at whatever cost is responsible for most of the ugliness, strife, and suffering that afflict the living community on the fairest of the planets illuminated by the Sun.

Despite this mad scramble to increase the quantity of organisms, often to the detriment of their quality, high values have emerged from the evolutionary process, most obviously beauty. Nature's beauty often makes us forget its darker side. From stately trees to delicate flowering herbs, lovely plants rarely repel us by their behavior. The characters of animals are more mixed; often we admire their grace or beauty while we abhor their conduct. Nevertheless, we find in the animal kingdom much to win our moral approbation: devoted nurture of offspring, mutual aid, cooperative societies, peaceful coexistence with organisms of other kinds, often mutually beneficial interactions between diverse species. Some of these behaviors are more difficult to explain by widely accepted evolutionary principles than are nature's harsher aspects. Evolutionists have used all their ingenuity to reconcile some of the most admirable developments in the living world with the prevailing concept of organisms eternally engaged in a relentless struggle to multiply their genes. Sometimes they have had to invoke special principles to prop up evolutionary orthodoxy.

To close our eyes to nature's darker side would be cowardly, dangerous, and unphilosophical; to dwell constantly upon it, depressing and alienating. I have long held that naturalists have no higher duty than to discover and make widely known all that is fair and heartening in nature, its values rather than its disvalues, for by this course we not only increase our love of our fellow creatures but support our own endeavor to overcome our weaknesses and approach more closely to our ideals. To know that gentleness, helpfulness, and friendly cooperation are not so rare in the natural world as they sometimes appear to be should raise our estimate of our prospects and fortify our determination to live in greater harmony not only with our human neighbors but also with the creatures of many kinds that surround us. The more violence and horror that we find in nature, the more imperative it becomes to contemplate everything good and encouraging that we can detect there.

High values are aesthetic, moral, and intellectual. Except possibly in their most rudimentary form, as in an animal's knowledge of its habitat, intellectual values appear to be confined to humans. Nature is an inexhaustible treasury of aesthetic values. To attribute moral values to nature is not to assert that any nonhuman animal has a self-conscious morality, although some may have more of it than we commonly suppose; they appear to have at least an incipient morality, or protomorality. When I ascribe moral values to nature, I mean no more than that we find among nonhuman creatures behavior that we can morally approve, behavior that conforms, at least objectively, to high ethical standards. And the harder we look, the more such behavior we find. In *Helpers at Birds' Nests* and other writings I have treated in detail certain of the moral values that the natural world presents. Although the present book deals primarily with aesthetic values, it will become evident that much of nature's beauty has been promoted by cooperation, a moral value.

A large part of the forms and colors that we find beautiful, including the blue sky, the green Earth, and the plants that adorn it (except their colorful flowers and fruits) would be just what they are if no eyes had ever beheld them — although, in a strict philosophical sense, they would not be beautiful without eyes to see and minds to appreciate them. To these we shall give only passing attention, while we concentrate on the beauty that would never have arisen in the absence of animals with vision. Even certain of the means whereby animals escape their enemies — concealing and warning coloration — have contributed to nature's beauty. To the mutually beneficial interactions of plants with the pollinators of their flowers and dispersers of their seeds we owe much of the loveliness of vegetation. The colors and adornments of many of the most beautiful animals, above all birds, have been promoted by preferential mating. Accordingly, the courtship of animals will receive a major share of our attention. Courtship assemblies, or leks, in which males gather to attract females, whose young they will not attend, have been outstandingly productive of beautiful plumage, as in birds of paradise, hummingbirds, and manakins. We shall look closely at these fascinating gatherings, in which cooperation and competition are subtly balanced. The bowerbirds of New Guinea and Australia, who for personal adornment substitute elaborate constructions and tastefully decorated gardens, are of such fundamental importance to the interpretation of avian courtship that we cannot neglect them. Then we shall see how mutual selection achieves for both sexes results not greatly different from those that intersexual selection, as practiced in courtship assemblies, brings to males released from parental tasks. After a briefer look at sexual selection in butterflies, we shall review the means whereby animals enmeshed in a process dominated by quantity bring quality into the realm of life.

Beauty is not confined to the land but abounds in the oceans, especially in the coral reefs, but to consider it in this book would make it too long, and lead me into a region where I lack experience.

Origins of Nature's Beauty

1.
Diverse Sources of Beauty

Nearly everywhere, at all times, a keen eye and alert mind finds beauty and interest in unspoiled nature, but in many parts of Earth certain months offer most to delight us. In the temperate zones, spring and early summer are the favored seasons. In the tropics, where rainfall rather than temperature controls the annual cycle of plants and animals, the beginning of the wet season corresponds to spring nearer the poles. Wilting vegetation renews growth; dormant seeds sprout; birds sing and nest; insects become more abundant. At these periods when nature is fairest and most exuberant, a leisurely walk through meadow and woodland, in bright sunshine, is a delightful experience. The more we pause to see and hear, the more carefully we observe, the more beauty we perceive — in the azure sky above us, the green Earth around us, the forms of leaf and flower, the colors of birds and butterflies and blossoms. Birds sing charmingly; flowers diffuse delicate fragrance. With so much of immediate interest to occupy our minds, we rarely ask how Earth came to be adorned with all this beauty. Nevertheless, understanding its origins greatly increases our enjoyment.

So diverse are the beautiful things around us that we hardly know how to sort them out into some semblance of logical order. A little reflection provides a clue. Some kinds of beauty arose independently of eyes to see them or ears to hear them; they were present on this planet long before any mind that might appreciate them; others would never have developed in the absence of eyes — or, in the case of sounds, ears — adequate to distinguish them. In this second category, the first to come to mind are insect- or bird-pollinated flowers, which evolved along with the bees, butterflies, birds, and other creatures capable of detecting and responding to their colors or scents. Similarly, the colors and aromas of fruits are closely related to the senses of the animals that eat them. The colors and adornments of animals have diverse functions. They may serve to attract mates endowed with color vision, to warn predators that they are distasteful or have adequate means of defense, to confuse pursuers, or to escape detection. Even deterrent colors are often not devoid of beauty. These are large topics to which we shall return in later chapters, while here we consider objects that were beautiful before animals developed eyes.

First and foremost among lovely visions is the azure sky which, when unpolluted, arches over a verdant landscape like a protecting dome. Of all things beautiful, it is the one we would most miss if we were destined never to gaze upon it again, the one we most welcome when, after days of gloom and storm, it is restored to us. Other beauty may become so commonplace that we scarcely notice it; it may even cloy the senses; but who can weary of beholding a wide expanse of the most tender blue, adorned with white clouds that are never twice the same, now delicate wisps and streamers, now long bars or banks, now the ribbed crests of aerial waves, now cumulus mountains towering high into the blue? As they float overhead, white clouds shelter us intermittently from the Sun's too ardent rays.

ₓ The other widely expanded source of beauty is the green Earth itself, whether covered with a dark mantle of forest, the paler green of meadows and thickets, or a mosaic of contrasting shades revealing different types of vegetation; whether lying flat over a wide plain, undulating over rolling hills, or more abruptly tilted in mountains. One who has long dwelt amid verdant nature misses the green of the Earth no less than the blue of the sky, and wonders how city people can remain contented in their absence. This green garment of the fertile land is the setting amid which we detect myriad forms of beauty.

Unlike much of the coloration of animals, the sky did not become blue nor Earth green in response to eyes that viewed them. They did not acquire their colors to please us. On the contrary, the comfort or delight that we find in them is probably an evolutionary adaptation. To those who dwell close to nature, a serene sky of the most soothing shades, spread over verdure that gives sustenance and protection, brings a dimly or keenly felt sense of well-being and security. These are the roof and floor of our true home. To find sky and Earth persistently ugly or forbidding would depress our minds and probably impair our health. Happiness might be impossible in such a setting. Not the least important aspect of our adaptation to our environment is our aesthetic adjustment to its most constant features, the colors of Earth and sky, our ability to find pleasure or comfort in their presence. They fortify our will to live. Although bats, owls, and other nocturnal creatures that drowse through the day in caves, hollow trees, or amid dense foliage might find a blue sky dappled with white clouds ugly or fearsome, such a reaction to them by diurnal animals would be disastrous.

Rainbows arched across the sky before ever there was a heart to leap up when beholding them, or even an eye to distinguish their delicately blended tints. Countless sunrises and sunsets colored sky and clouds before there were creatures to notice them. Drops of rain or dew, hanging from leaf or stem, scintillated brightly when swayed by gentle breezes in morning sunshine. Mountain torrents sparkled and foamed where sunshine struck them. After nightfall, myriad stars shone out in the dark firmament, long ages before minds were stimulated by this vision of splendor and vastitude to reflect upon the immensity and mystery of the Universe. The moon, waxing and waning in its monthly cycle, shed its soft light upon its mother planet with never a spirit sensitive to its charm. Animals can only passively enjoy the sublimity of the heavens; even humanity can hardly intensify it, although we can dim it by polluting the atmosphere.

Crystals of many kinds formed without eyes to see them. For ages precious gems — diamonds and rubies and sapphires — lay embedded in the rocks, as though waiting for humans to extract them, to be fascinated by their hardness and glitter, and, unfortunately, to covet them, too often inordinately. Atmospheric water crystallized in countless hexagonal patterns, often of great complexity and charming delicacy, to fall as snowflakes upon Earth still devoid of creatures able to appreciate their fragile loveliness. In short, lifeless matter spontaneously assumes innumerable harmonious configurations wholly uninfluenced by the living world. Aeons passed before humanity learned to employ it in all the productions of art, which, however beautiful, never attain the grandeur of nature's creations.

When we turn from the lifeless to the living world, we find both categories of beautiful things. The appearance, especially the coloration, of numerous animals and even of vegetable organs is a result of the combined action of blind vital processes and selection by creatures with vision; but many other organic forms and colors would be just as they are in the absence of eyes that see them. The endlessly varied shapes of leaves, simple or of great complexity like the filigree fronds of tree ferns, are expressions of their formative processes as modified by the environment, but hardly influenced by animals, except in the case of the minority that may have become thorny or prickly to resist browsing or grazing animals. The stately forms of trees have not been shaped by animals that alight upon or climb through their boughs so much as by meteorological factors, such as snow, which appears to be responsible for the tapering form whereby northern conifers distribute its weight among branches that become progressively shorter toward the top.

The green of vegetation is the color of chlorophyll, which by photosynthesis supports all the planet's life except the minute fraction dependent upon chemosynthesis by tiny organisms. Plants contained chlorophyll before animals learned to profit by its activity. The brilliant autumnal tints of deciduous broad-leaved trees in the North Temperate Zone are caused largely by anthocyanin pigments that become prominent as chlorophyll fades; they are incidental rather than integral to the process of defoliation and appear to be without significance to woodland animals. It is not so widely known that falling leaves of many tropical trees are no less highly colored — often yellow, orange, or red — than those of temperate-zone trees. Usually they drop one by one from a tree that remains green and perhaps simultaneously expands new foliage, so that in humid tropical regions dying leaves never color a whole forest and rarely give the prevailing hue to the tree that drops them. Young leaves of tropical trees are often beautifully tinted with pink, purple, or bronze while they hang limply until they attain full size, their tissues harden, and they rise to a more horizontal position, the better to intercept the light.

In contrast to the vegetative organs of plants, whose forms and colors are rarely modified by interactions with animals (however much animals may distort them by browsing, grazing, and gnawing), floral shapes, colors, and scents have evolved in relation to pollinators, as the colors and structures of edible fruits have done in relation to the dispersers of their seeds. These are subjects to which we shall later return.

The forms of animals are determined by their modes of locomotion and foraging and the media in which they live. The streamlined bodies of fishes, dolphins, and other aquatic creatures facilitate their movement through water. The slender grace of antelopes, like the shapely bodies of horses and zebras, are the foundation of their fleetness. Feathers, wings, and shapes adapted for efficient flight give elegance to birds, apart from their frequently lovely colors. Although bats fly well, their wings devoid of plumage fail to make them beautiful. Some animals are most attractive when young; others, when mature. Newborn colts have spindly legs that seem too long for their bodies; as their bodies fill out, their appearance improves. With horned cattle, the reverse is true; calves and heifers are more comely than mature cows.

The colors of animals are largely, but by no means wholly, determined by their interactions with other animals that can see them — a large subject that will claim our attention in most of the following chapters. Much depends upon the texture of their coverings; hair does not lend itself to bright coloration as well as feathers, scales, or bare skin do. Climate affects the colors of birds and mammals, as recognized by Golger's Rule, which states that races inhabiting warm and humid regions tend to be darker than those of cool and arid areas. These differences in shades may be related directly to the physical environment. Dark colors absorb radiant energy more freely than light colors; one of the problems of denizens of hot deserts is to avoid overheating; their paleness may help them to keep cool. On the other hand, their light coloration also makes them less conspicuous against the sand or pale soils of arid regions, in which case it is promoted by interactions with other animals, especially predators. The effects of the two factors are difficult to untangle and doubtless differ with the species.

The beauty of organisms is not confined to their external forms and exposed surfaces. The nacreous luster of the inner surface of the shells of certain mollusks is an incidental effect of their minute structure, devoid of survival value, not meant to be seen by any eye, but nevertheless highly attractive. Viewed through a microscope, many vegetable tissues, composed of cells arranged with great regularity, are pleasing to behold. Although much of nature's beauty has arisen in relation to eyes that see it, all that we detect in the physical realm, and no small part of that of living organisms, might exist in a sightless world.

2.
Beauty and the Aesthetic Sense

For simplicity, I wrote in the preceding chapter of beauty in a sightless world. Now I must admit that such a world would be devoid of visual beauty, however richly it might be endowed with forms and colors. To understand this paradox, we must look at the nature of values, a large category in which beauty is prominent.

Philosophers apply the term "value" to whatever gives pleasure, embellishes life, or enhances existence. From ancient times, beauty, goodness, and truth have been regarded as the three principal categories of high values. Beauty is value perceived directly by the senses, especially sight and hearing; goodness is moral value; truth is value in the realm of intellect. All are modes of harmony; the first, between sights and sounds and a perceptive spirit; the second, in our daily lives and relations with the creatures around us; the third, among the ideas in a mind. Moreover, they overlap: beauty and truth are certainly good; goodness and truth are beautiful. Socrates, in Plato's *Phaedrus,* prayed that he might be made beautiful in the inner man; Shelley wrote a hymn to intellectual beauty, "the awful shadow of an unseen Power."

Values may be of internal or external origin. To reach, by thought alone, a neat solution of a perplexing problem is a high value with no immediate external source. Except perhaps for philosophers, mathematicians, and others deeply involved in intellectual pursuits, most high values are of external origin, as is certainly true of aesthetic pleasures. They are not something that we pluck ready-made from a beautiful object, let us say a flower, nor are they, as Platonists hold, essences eternally present in a changeless realm. Every value is a fresh creation, born of the interaction of an appropriate object, the value-generator, and a receptive mind, the value-enjoyer. The value-generator may be a flower, a rainbow, a bird's song, a cheerful greeting, a helping hand; the value-enjoyer, an adult human, a child, perhaps an animal. Unless we assume that lifeless matter is neither wholly insentient nor devoid of the possibility of enjoyment however slight, values cannot arise in a lifeless world. In any case, the blue sky and green Earth, rainbows, sunrises and sunsets, refulgent gems, and similar objects lacked the kinds of values that we find in them, lacked beauty, until beings with aesthetic sensibility appeared on Earth. Only potentially beautiful in the absence of value-enjoyers, these things waited a long age for their beauty to become actual.

A value such as beauty is, as we have seen, born of the union of two distinct entities, a value-generator and a value-enjoyer. We might think of the value-generator as the father and the value-enjoyer as the mother. As in the birth of a child, the mother contributes more than the father. The latter initiates the process of embryonic development and, at least in part, determines the form of the new life, but the mother nourishes and protects it. A value-generator such as a flower is a complex structure which in sunshine emits a delicate fragrance and diffuses in all directions a selection of the light waves that fall upon it. A minute fraction of these intrinsically colorless waves enter the pupils of the beholder's eyes and are focused upon the sensitive retinas. By a process that we are far from understanding in detail, the separate patterns on the retinas are united in a single colored image in the mind of the observer, who, if in the proper mood, is delighted by this vision. Since perception of the flower requires organic structures more complex than the flower itself and, moreover, its colors are added by the percipient, it is no exaggeration to say that the observer contributes more than the flower does to the value that he or she enjoys. However, since without the value-generator the pleasant experience would not have occurred, the observer spontaneously objectifies the value, projecting it outward upon the thing that delights, as though in grateful recognition of what is owed to it.

Like delicate flowers, values bloom and fade. A second view of the same beautiful object, a second audition of the same melody, is not the same value repeated but a wholly new value which, however much it might resemble the first, is numerically distinct. Some values are slight and evanescent, soon forgotten; many leave faint or vivid traces in memory. If it has no stubbornly persistent source of sorrow, a life rich in values, of which one of the greatest is health, is likely to be happy. Thus, happiness might be regarded as a summation of values, or perhaps as the supreme value, to which all other positive values contribute.

Beauty is not an instrumental value, like a bitter medicine that one takes to relieve pain, but an intrinsic value, welcomed for itself. Rarely are other values so swiftly and uncritically appreciated. Our response to bright colors, shapely forms, and pleasant sounds tends to be immediate and direct. This is undoubtedly because beauty is primarily a sensuous experience, and the minds of animals have evolved to appraise and respond rapidly to the reports of their external senses; their lives may depend upon the swiftness of this response. Nevertheless, an intellectual element is rarely absent from contemplation of the highest beauty, especially if the pattern is intricate or the form complex. We examine the details of a natural object or a work of art, study the relations of its parts, appraise the balance of the whole, and try to fathom its significance. Is the object beautiful or merely gaudy? Has the artist accomplished what he tried to do? This critical assessment may increase our enjoyment of the object by revealing harmonies not apparent at the first glimpse or may spoil our pleasure in it by revealing obscure defects or offensive details. The face that at first attracts by its youthful freshness may, with growing familiarity, repel by the coarseness of its features or the unpleasant lines of the mouth. The painting that pleases by its bold conception and rich colors may reveal poor craftsmanship that alienates a critical eye.

Our appreciation of an object with claims to beauty can hardly fail to be strongly affected by its associations, or what in aesthetic theory is called its expression. Although our response to beauty tends to be immediate and unquestioning, it is modified by knowledge of the beautiful object's character. We readily find aesthetic defects in that which repels us for practical or moral reasons. A colorful fruit becomes less attractive when we learn that it is poisonous. Nothing so detracts from an animal's appearance as a long jaw with prominent fangs. Where we see beauty, as in a lovely face, we spontaneously expect goodness; to learn that it is lacking disappoints and estranges us — which seems to impose upon the person whom nature has made physically beautiful the obligation to become morally beautiful as well. And just as revolting associations can detract from beauty, so can pleasant associations enhance it. Since, by a minimum definition, beauty is pleasure springing up immediately upon the sight of an object, a well-loved face although old and wrinkled, a house where we have dwelt long and happily although slightly dilapidated, a favorite picture that has faded are often beautiful in the sight of one who loves them, no matter how homely they might appear to another.

The materials of visual beauty are form, color, pattern, and texture. The most pleasing forms have either radial symmetry, as in many flowers and marine invertebrates such as starfishes, or bilateral symmetry, as in other flowers and nearly all vertebrate animals. The forms of organisms, especially active animals, are determined largely by their functions and, accordingly, are closely linked to utility. The prevailingly rounded rather than angular outlines of well-nourished, healthy animals make them more pleasing; but obesity is unattractive. To watch forms in themselves pleasing engaged in the activities for which they evolved increases our delight in them, as in galloping horses, birds soaring on wide white wings, dolphins racing in front of an advancing ship. Even monkeys, not the most graceful of animals, win admiration by their prodigious leaps across gaps in the forest canopy, for which arms disproportionately long for their bodies so well prepare them. Hummingbirds are loveliest when they hover motionless on wings beaten into a halolike haze, sipping nectar from flowers, with their glittering, iridescent gorgets turned squarely toward us. Most small birds are best seen at rest, when we can enjoy their forms and colors; in flight, details of plumage that are their chief attraction are lost to us. In architecture, form cannot be divorced from utility. A building not adequate for its purposes fails to please, nor does one that does not appear to be substantial, no matter how well the architect can demonstrate that the seemingly too slender supports can bear the structure's weight.

Color is subordinate to form. Daubed at random on canvas or a wall, the brightest colors fail to please; but a graceful form devoid of color, a statue in pure white marble, or a design in black and white is often beautiful. Most colors and many neutral shades are attractive in an appropriate context. Exceptions are colors, especially toward the red end of the spectrum, so intense and harsh that they seem to shriek, those lacking in purity, and shades of gray and brown that may be more depressing than black, which is the absence of all color. Why certain colors repel us, even in the absence of unpleasant associations, is an interesting question difficult to answer. The analogy of sounds provides a hint. May it not be that, just as a mixture of sound waves of discordant frequencies makes a distressing din, so a mixture of light waves of incompatible frequencies produces a disagreeable color?

Pattern or design refers to the distribution of shapes and colors over a form or surface. In general, simple patterns that the eye can follow are more pleasing than intricate designs that dazzle vision; Grecian simplicity is for many of us more beautiful than arabesque profusion. Nevertheless, the total effect of an intricate pattern, which often consists of innumerable, crowded repetitions of the same figure, as on many fabrics and the plumage of certain birds such as pheasants, may be impressive, suggesting richness or profusion and, on a product of human handicraft, of prolonged, patient application that commands admiration. A majestic tree is covered with countless leaves that vary only slightly in shape; the innumerable visible stars that adorn the nocturnal sky hardly differ except in brilliance.

Texture denotes the character of a surface that bears color, making it cold and glittering or soft and warm. The texture of feathers contributes greatly to the beauty of birds, whatever color they bear. Although the naked heads of certain birds, such as vultures, may be repulsive, their feathered parts are nearly always attractive; even black plumage, whether glossy or velvety, is frequently pleasing. Like the feathers of birds, the scaly wings of butterflies and moths have a texture that enhances their colors. How different the soft coloration of butterflies' wings from the metallic glitter of the shards of many beetles! On surfaces so extensive that we fail to notice form or its absence, texture devoid of pattern can be beautiful, as on deeply colored draperies of velvet or some other rich material, the azure sky (which we see as a surface, although it is not exactly that), and the ultramarine of a tropical ocean spreading calmly in brilliant sunshine. Nevertheless, variety and contrast intensify beauty and prolong our enjoyment of it, whereas monotony fails to hold interest. Whether in the productions of nature or of art, these four — form, color, pattern, and texture — are the elements of which beauty is compounded.

Beauty, like other values, has degrees, which we express by such adjectives as pretty, beautiful, exquisite, and magnificent. One responsive to beauty finds few places on Earth's surface where it is wholly absent. No matter how harsh or enervating the climate, how forbidding the terrain, beauty awaits the sensitive eye, in the colorful, wind-sculptured rocks and brilliant sunsets of the desert, in the gleaming snow and spectacular meteorological phenomena of arctic regions, in myriad forms and colors of plants and animals in hot tropical forests. The strong contrasts of the objects and scenes that we find beautiful make us ask in what beauty consists and why we are so sensitive to it. The explanation appears to be our pleasure in the exercise of our externally directed senses, especially vision, the most precious and useful of them. Our minds spontaneously and tenaciously strive to recognize form, pattern, and meaning in whatever attracts our attention and holds our vision; the success of this endeavor is satisfying or pleasurable. We find beauty nearly everywhere because it is our nature to find it, as it is our nature to breathe air. Our aesthetic adaptation to Earth's diverse regions is broader and more flexible, more perfect, than our physiological adaptation. The beauty of many a desolate land, hostile to human life, draws adventurous spirits to visit and revisit it, despite hardships and perils. Beauty binds us to Earth, in all its contrasts and extremes, as nothing else can.

When we contemplate the vast diversity of things that we call pretty or beautiful, the only feature that we find common to all is their capacity to stir a pleasant, comforting, or, at strongest, joyous sense of their presence around us. Or, as I wrote elsewhere, beauty is our delighted awareness that other beings coexist with us.

Of the three traditional high values, beauty is the most widespread and primitive. That beauty abounds in the living world admits no doubt, but the degree to which nonhuman creatures appreciate beauty — make a value of it — is far from clear. Lacking eyes, ears, and a central sensorium, plants cannot be aware of their own loveliness or of the melodies of birds that sing among them. Some animals, above all birds, give strong indications that, in respect of beauty, they are value-enjoyers as well as value-generators, that they are sensitive to beauty — a matter to which we shall return in later chapters.

At an early age, before they develop a conscience or seek truth, children reveal a nascent appreciation of beauty by their responsiveness to colorful and glittering baubles. Even monkeys and apes enjoy daubing paints on paper or whatever surface is available to them. Among primates, aesthetic sensibility has raced ahead of other valuable attributes, such as moral responsibility, moderation, restraint, compassion, and the capacity to care devotedly for whatever supports or delights them. In one respect, this precocity of the aesthetic sense is beneficial, for attraction to beautiful things may eventually lead us to protect and try to understand them. But in conjunction with widespread acquisitiveness and destructiveness, it has been disastrous. Apes rudely tear apart things that strongly attract their attention. From ancient times, conquering armies have borne away, among the spoils of war, the fairest art treasures of a vanquished city. Children's eagerness to pluck lovely flowers, men's readiness to shoot and stuff beautiful birds have brought species of plants and animals to the verge of extinction — or beyond. The flower wilts; the dead bird loses its charm; neither reproduces its kind; beauty is lost. The beauty that attracts us to plants and animals has, all too often, been a curse to them — as beauty has been to many a comely person. Beauty should be contemplated, not physically possessed. When thoughtless acquisitiveness is overcome, beauty brings added benefits by promoting the growth of its kindred values, goodness and truth. Reflecting that the beautiful thing contributes to our enjoyment simply by being itself and expressing its own nature, we wish to avoid injuring it, to permit it to remain itself; we desire to live in harmony with it, which is goodness. Often, too, we try to learn more about it, to understand it more deeply, thereby increasing knowledge or truth.

Finally, we wish to know why beauty delights us, why we are blessed with aesthetic sensibility. It is not immediately evident why an animal, human or other, should be interested in, and often deeply stirred by, so many things that contribute nothing to individual or racial survival — neither food, nor safety, nor reproduction. Our fashionable evolutionary theories fail to elucidate this question; they account for nothing that does not promote the multiplication of organisms, often excessively. To understand why we love beauty, we must consider what we essentially are.

Whatever else a man or woman or any other organism might be, all are products of the universal process of harmonization, which builds up patterns of increasing amplitude, coherence, and complexity. On the small scale, it joins the elementary particles in atoms and molecules, which in appropriate conditions line up, rank upon rank, in crystals, our first evidence that harmonization creates beauty. On the largest scale, it condenses great masses of matter into suns, planets, and moons, and sets them revolving around one another in systems so balanced and stable that they endure for long ages, as in the solar system, the grandest example of an integrated pattern that we know. In the living world, harmonization is manifest as growth, in which molecules combine to form protoplasm and cells, cells proliferate into tissues, and tissues compose organs, which together form the organism.

The parts of a plant — root, stem, leaf, and flower — are bound into mutual dependence by the circulation of sap and elaborated materials, by hormones, and rather loosely by protoplasmic strands that penetrate cellulose walls from cell to cell. The body of one of the more advanced animals, especially a vertebrate, is far more tightly integrated by the circulation of blood, by hormones carried to all parts by the circulatory system, and by a nervous system reporting to, and bearing commands from, a central sensorium or brain. The animal's health depends upon the harmonious cooperation of its organs and vital functions, and equally upon harmonious adjustment to its environment. A product of harmonization, its welfare depends upon the maintenance of harmony. Its primary nature is determined by the process that created it.

In many animals, this primary nature is overlaid and masked by passions and behaviors foisted upon it by evolution in the struggle to exist in a world made fiercely competitive by life's unrestrained fecundity. They are burdened with offensive and defensive weapons or clad in protective armor. They become savagely aggressive or timidly unapproachable, or each in turn, as occasion demands. Nevertheless, as long as it remains healthy, an animal's internal activities remain true to the process that formed it, however antithetic to this process its external activities may become.

In the long and checkered course of its evolution, humankind has been infected, often with greater intensity, with the same disruptive and distressing passions that afflict other animals. In the measure that humans can mitigate these passions, their primary nature asserts itself. Products of harmonization, and throughout their active lives participating in a dynamic process of harmonization, they strive to preserve internal harmony, which is health, and external harmony, or adjustment to the environment, without which internal harmony deteriorates. However artificial they may make their environment, harmonious adjustment to it is no less vital to them than adjustment to its natural ambience is indispensable to any other organism, vegetable or animal.

Moreover, in the measure of our psychic or spiritual development, we find pleasure in, and try to cultivate, all felt or perceived harmonies — our felicity depends upon success in this endeavor. We try to make thoughts, words, and deeds conform to our guiding principles, whether we were taught them in childhood or developed them for ourselves; a relatively untroubled conscience is our reward for achieving this conformity. To the best of our ability, we cultivate harmony with the creatures around us, both human and nonhuman, but success in this endeavor depends upon their cooperation, which, unhappily, is not always forthcoming. As Pittacus proclaimed of old, it is hard to be good. We delight in knowledge, mental clarity, and coherence among the ideas in our minds, which is our most reliable criterion of truth. And we enjoy beauty, which we perceive when visible objects or sounds harmonize with our sensory and psychic constitution. We attribute our delight in beauty to our aesthetic sense, a convenient designation which becomes misleading when we take it to denote an isolated psychic faculty.

Our capacity to recognize and enjoy harmonies revealed by our distance receptors, especially vision and hearing, is but a particular instance of our responsiveness to harmony in general, rooted in the very core of our being. Our quest of truth, our efforts to be good, our concern for our health, are diverse directions taken by the same fundamental preference for harmony. So closely related are they that none can ignore the others without weakening itself. Life degenerates, becoming feeble, unbalanced, or absurd, when beauty, knowledge, goodness, or health is cultivated without due regard for its sister values. Losing sight of this truth, the doctrine of "art for art's sake" is responsible for a vast clutter

3.
Beauty That Conceals or Repels

That the adornments which animals display to attract or impress others of their kind, especially mates, frequently appear beautiful to us should not surprise us; these animals are usually endowed with color vision and presumably see things much as we do, although often more acutely. But to find attractive patterns that serve their purpose best when they remain unseen is so unexpected that we ask why designs so intricate as some of them are should adorn creatures that persistently try to hide them. And when the function of an animal's colors is to repel predators rather than to attract partners, should they not be ugly and fearsome rather than beautiful? Nevertheless, we often find warning coloration far from displeasing. Beautiful animal coloration that sometimes appears to be functionless or misplaced will occupy our attention in this chapter.

Concealing (procryptic) coloration is widespread among birds who pass much of their lives on the ground, breeding there in shallow, open nests, or none at all. Such a bird is the Common Pauraque (*Nyctidromus albicollis*), a nightjar that ranges from southern Texas to northern Argentina. All day it rests upon the ground in the shade of open woods or thickets, on brown fallen leaves and litter with which its plumage blends so well that if it slept soundly a person might inadvertently tread upon it. However, the bird remains alert even by day, rises lightly in front of the intruder, circles around, and alights in another spot. Seeing the bird fly up, one would hesitate to call it beautiful. Without the least attempt to build a nest, the pauraque lays two pale buff or pinkish eggs directly on the ground, usually in a shady or somewhat open spot, where from a blind I have spent long hours watching both sexes incubate, alternately. The longer I gazed upon the intricate pattern of delicately blended shades of brown, gray, buff, and black of their soft plumage, the more beautiful these birds appeared to me, until I was convinced that they were no less lovely than many a bird so brilliantly attired that at the first glimpse one exclaims "How beautiful!" But their beauty is of a different order, that must be contemplated at leisure, and at close range, to be appreciated. The sexes of this nightjar are nearly alike, except that the white areas on the male's wings and tail are larger and purer than those on the female. Pauraques do not become active until daylight fades, and even bright moonlight would hardly reveal the pattern on their dark plumage, making it probable that at most the white patches play a part in courtship.

Other nightjars, many shorebirds, snipes, sandgrouse, female ducks, and sparrows, to mention only a few, similarly clad in blended neutral shades, are also beautiful to the contemplative eye. Very different is the concealing coloration of birds of leafy treetops. The green of many kinds of parrots matches the foliage of a spreading tree so well that a large flock of them may rest or forage there, unnoticed until they burst forth with a raucous din, revealing, as they fly away, patches of red, yellow, or blue that make them more handsome than concealing coloration requires. The green of shrike-vireos (*Smaragdolanius* spp.) is brighter than that of the forest canopy, where they are exasperatingly difficult to see even when they proclaim their presence with a triple whistle tirelessly repeated. Many small birds that forage on leafy boughs are attired in soft, pleasing shades of olive, pale yellow, or gray.

The coloration of butterflies and moths is related not only to their periods of activity but perhaps more to the way they hold their wings when at rest. Diurnal butterflies fold their wings above their backs, dorsal surfaces together. Nocturnal moths rest by day with wings spread broadly against the supporting surface, forewings usually covering the smaller hindwings. The dorsal surface of the moth's forewings is usually colored to match the bark or rock where the insect rests, immobile and difficult to detect. The grays, browns, and buffs of these wings are frequently arranged in a pattern more elaborate than would be needed for camouflage.

Among the more beautiful of the large moths is the Cynthia (*Samia cynthia*), whose wide wings are softly tinted with gray, brown, pink, orange, lavender, and white, with an eyespot at the tip of each forewing. Another handsome species is the sphinx moth *Deilephila hypothous* of New Guinea, more deeply colored with brown and velvety black, with an eyespot near the base of each forewing. A broad light band extending transversely across the wings and thorax of this sphinx moth breaks its form into two parts and makes it more difficult to recognize.

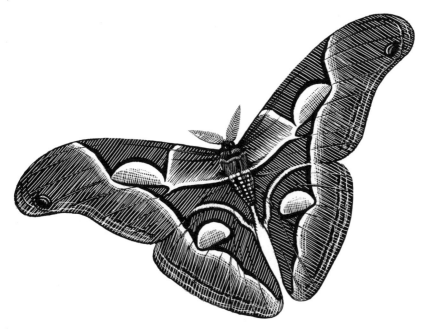

Cynthia Moth, *Samia cynthia*.

Other creatures confuse their enemies by sudden changes in appearance. In his wide experience as a traveler and collector in the tropics of both hemispheres, Alfred Russel Wallace knew no more "wonderful and undoubted case of protective resemblance in a butterfly" than that of the common Indian *Kallima inachis* and its Malayan ally, *K. paralekta*. These large leaf butterflies commonly rest in a bush or tree in dry forests, amid dead leaves which they closely resemble. Their wings are tightly folded above their backs, exposing the lower surfaces, which vary from individual to individual but are always ashy or brown or ochre, often with holes and blotches and spots that simulate foliage in all stages of decay. Between the acute apex of the forewing and the short, narrow tail of the hindwing run lines that suggest the midrib and lateral veins of a leaf. The wings' irregular outlines strengthen their resemblance to withered foliage. Simulating a petiole, the tails of the hindwings touch the supporting twig, to which the butterfly clings with its inconspicuous middle pair of feet. Its head and antennae are hidden between the closely appressed bases of the wings. If this motionless dead leaf is disturbed, it springs into life as a butterfly whose broad wings are adorned on the upper surface with a wide band of rich orange on a deep bluish or purplish ground. It flies swiftly for twenty to fifty yards, then vanishes as it alights in an appropriate spot and folds its wings together. So sudden is the change in appearance and so perfect its camouflage that, unless one had noticed exactly where it has come to rest, it is exceedingly difficult to find.

Among North American butterflies, the angle-wings perform similar disappearing acts. The large Violet-tip, or Question-mark (*Polygonia interrogationis*), has the upper surface of the wings beautifully marked with dark brown upon a ground of orange-brown, with a wonderful violet iridescence playing over all in bright sunshine. The wings' undersurfaces are brownish gray, indistinctly marked, so that when the butterfly alights and folds them above its back it resembles the bark of a tree. Morpho butterflies flash wide expanses of the most intense azure as, with swift and erratic flight, they thread the lower levels of tropical American forests. When they alight and fold their wings, exposing only the dark undersides, they suddenly vanish. However, they appear to have few enemies — among birds, chiefly jacamars — for they are not always careful to choose an appropriate background.

Among moths, the underwings of the genus *Catocala* change their appearance confusingly as they take flight or come to rest. By day they cling to trees, exposing little more than the mottled grayish and brown surface of their spread forewings, which assimilates them to the bark. When molested they fly, flashing bold patterns of bright colors on their hindwings, which vanish as they alight on another tree, to the confusion of the bird whose eyes are fixed upon an insect of quite different aspect. The hindwings of the Io Moth (*Automeris io*), each with a large eyespot set amid soft tints of orange, yellow, and pink, are covered with much duller forewings while the insect rests. The upper surfaces of forewings and hindwings of the Eyed Hawk Moth (*Smerinthus ocellatus*) of Europe show similar contrasts. Since moths are active by night, and in any case moving objects attract attention whatever their color, the patterns of their hindwings probably do not compromise their safety, which by day is promoted by fleeting exposure of beauty that suddenly disappears when a disturbed insect alights.

Certain grasshoppers baffle their pursuers by the flash-and-cover trick widely employed by butterflies and moths. Their forewings, green or gray or variegated to match the grass, soil, or rock on which they rest, cover more brightly colored hindwings, black bordered with yellowish white on the Carolina Locust (*Dissosteira carolina*) of eastern North America; red bordered with black on the coral-winged locusts; crimson spotted with blue-black on the giant Red-winged Grasshopper (*Tropi-*

dacris cristata) of tropical America, of which the adult female is nearly six inches (fifteen centimeters) long, with a wing span of about ten inches (twenty-five centimeters). Although the adults of these big grasshoppers are conspicuously colorful only when they spread their wings in flight, their immature stages are boldly striped with reddish brown, an almost certain indication that they are distasteful to insectivores.

Although creatures eagerly devoured by predators are often concealingly colored and try to remain undetected or, if discovered, to confuse their pursuers by abruptly changing their appearance, those protected by bad taste or irritating chemicals find it advantageous to advertise their immunity by bright colors or bold patterns. Among organisms with warning (aposematic) colorations are many caterpillars, of which those of the widespread Monarch butterfly (*Danaus plexippus*) are probably the most familiar in North America. In their final instar or larval stage, the fat body is encircled by bold alternating bands of black, cream, and yellow, making these caterpillars conspicuous as they gnaw leaves of milkweeds (*Asclepias* spp.) with little attempt to hide themselves, as though confident that their taste would protect them from caterpillar-eaters. The caterpillar of the Eastern Black Swallowtail (*Papilio polyxenes*), widely distributed in North America south of Canada, is no less conspicuous than that of the Monarch. Its green body is encircled by black bands that are alternately broad and narrow, and the former are marked with orange-yellow spots in a regular and attractive pattern. This caterpillar eats the foliage of carrots, parsnips, parsley, and other umbelliferous plants. When disturbed, it thrusts out from just behind its head an orange-yellow, Y-shaped organ, known as an osmeterium, which emits a most unpleasant odor that appears to deter hungry birds.

In his book *On Natural Selection,* Wallace records observations, made by various British naturalists in the middle of the nineteenth century, on a number of smooth, conspicuous caterpillars. The white-and-black-spotted larvae of the Magpie Moth (*Abraxas grossulariata*) and the yellow, black-spotted caterpillars of the Six-spot Burnet Moth (*Zygaena filipendulae*) were consistently rejected by birds, lizards, frogs, and even spiders. If seized by a lizard or frog, they were immediately dropped with evident disgust, and thereafter ignored. Similarly, the caterpillars of the Figure-of-eight Moth (*Diloba caeruleocephala*), pale yellow with a broad blue or green lateral band, and those of the Mullein Moth (*Cucullia verbasci*), greenish white with yellow bands and black spots, were ignored by a variety of birds that eagerly devoured unprotected caterpillars offered to them at the same time.

More recently, Niko Tinbergen reported observations on the palatability of caterpillars of the Cinnabar Moth (*Euchelia jacobaeae*), which when half- or full-grown are extremely conspicuous with alternating black and yellow rings. Moreover, they live in groups on plants of the Ragwort (*Senecio jacobaea*), often completely defoliating them. Young, inexperienced birds do not hesitate to seize these boldly attired caterpillars, only to drop them with signs of disgust, then vigorously wipe their bills as if to remove a bad taste. Thereafter, the birds cannot be induced to touch these Cinnabar Moth larvae. The adult stages of some of the foregoing moths are both colorful and unpalatable.

Adult moths with warning coloration include tiger moths of the family Arctiidae, whose wings are brightly colored in elaborate patterns, often of great beauty. A European species inspired John Keats to write:

> All diamonded with panes of quaint device,
> Innumerable of stains, and splendid dyes,
> As are the Tiger Moth's deep damask wings.

The bright red of ladybug beetles, boldly spotted or otherwise marked with black, warns hungry birds that they are not good to eat. True bugs (Hemiptera), many of which are protected by bad odor as well as bad taste, are often brightly colored, frequently red. One of them, a cotton stainer (*Dysdercus* sp.) is attractively adorned with red, blue, and buff, with a large black spot on each forewing.

On dimly lighted ground and on the lowest undergrowth of rain forests in the Caribbean lowlands of Costa Rica lives a tiny frog (*Dendrobates pumilo*), about three-quarters of an inch (two centimeters) long, which on wet days in May calls attention to itself by puffing out its throat and, with vibrating abdomen, emitting an insectlike buzz. With brilliant orange-red bodies and deep blue legs, all finely spotted with black, these abundant frogs are not hard to find. By voice and coloration, they appear to try to call attention to themselves and, unlike less brilliant frogs, are in no hurry to escape when approached. The chickens and ducks to whom Thomas Belt offered this or a similar species in Nicaragua would not touch it. When, by a ruse, he induced an inexperienced duckling to seize one of these little frogs, the poor bird instantly dropped it, and went about jerking its head as though trying to rid itself of something disagreeable.

Another frog with brilliant warning colors is the slightly larger *Atelopus varius,* which is black and green and, on the larger individuals, red. It often rests on rocks in and beside clear streams in forested Costa Rican foothills, where it is so sluggish and reluctant to move that it owes its life more to one's care to avoid stepping on it than to any prudent effort of its own. Some of these small, colorful frogs of tropical America are well known for the virulence of their toxins, which Amerindians have used to poison their arrowheads. Among reptiles, outstanding examples of warning coloration are the venomous coral snakes, among the most brilliant of serpents, whose coral red bodies are ringed with black and yellow. They are closely mimicked by snakes harmless to humans.

No other insects contribute as much to nature's beauty as butterflies, whose wide, colorful wings display an endless variety of intricate patterns. Butterflies could not be so conspicuously abundant if they were not, on the whole, avoided by the most numerous and active of diurnal insectivores, the birds. The size and stiffness of their wings appear to make these insects unattractive to birds, who, before swallowing the bodies, usually remove them by vigorously beating their victim against a perch. While a bird is so engaged, the wings flap in its face in a way that can hardly be pleasant. Where other insects that are more easily prepared for eating are plentiful, most birds prefer them to butterflies. None that I know takes more large butterflies than jacamars, whose long, thin bills can reach past the wings to seize the insect's body and hold it away from the jacamar's face while the bird removes the wings.

Jacamars do not take all butterflies indiscriminately. In a Venezuelan ravine where heliconians were amazingly abundant, a pair of Pale-headed Jacamars (*Brachygalba goeringi*) ignored these slender-bodied, long-winged butterflies while they brought many skippers, along with dragonflies and other insects, to their four nestlings in a burrow in a bank. Butterflies do not depend wholly on the breadth of their wings to discourage predators. Like the heliconians, many are distasteful to birds. Although the Monarch is generally considered to be a protected species, a leading investigator of this familiar butterfly, F. A. Urquhart, sampled it and found it tasteless. Conflicting reports of the acceptability to birds of Monarch butterflies were reconciled by L. P. Brower's experiments with Blue Jays (*Cyanocitta cristata*). In these butterflies, as apparently in many other insects, the chemicals that make them unacceptable to birds are not synthesized by themselves but derived directly from the plants they eat in the larval stage. Jays who ate Monarchs raised on certain species of milkweeds were made violently ill by cardenolide heart poisons that they contain, and after recovery rejected similar butterflies. Monarchs nourished by species of *Asclepias* devoid of toxins were consumed with impunity by jays.

Plants probably could not cover Earth with verdure if they lacked defenses against the numerous herbivores, including insects in their larval and adult stages. These defenses include spines, stiff irritating hairs, thick cuticles, or the like, but appear to be most often chemicals injurious to plant-eaters. In the course of evolution, many insects have become immune to poisons present in a certain species, genus, or family of plants; just as, in recent decades, they have developed resistance to the pesticides that humanity pours over the environment. Whereas some insects appear to detoxify the poisons in the vegetable tissues that they eat, others retain them more or less unaltered, thereby gaining a double benefit. Not only can they eat certain plants repellent to other insects; they become distasteful or dangerous to animals that would eat them, and it is to their advantage to advertise this fact by developing warning coloration.

Many warningly colored insects are not merely unpalatable but contain chemical compounds capable of causing physical pain, distress, or fright to birds or other predators. When an aposematic insect was placed in the cage with a bird that had already had a bad experience with it, the bird displayed or uttered alarm notes. If the offending insect was left in the cage, the bird tried frantically to escape. Birds lack innate recognition of the many protected insects of diverse appearance found in their habitats but must learn to avoid them. Miriam Rothschild believed that by watching the signs of distress of an inexperienced companion who first tries to eat a certain unpalatable or poisonous insect, a bird may thenceforth eschew this insect without having to learn the hard way.

Palatable insects are often confusingly similar to unpalatable insects in the same habitat. Thus, the palatable Viceroy Butterfly (*Basilarchia archippus*) closely resembles the usually unpalatable Monarch; and many swallowtails (*Papilio* spp.) resemble other swallowtails or butterflies of different genera that are distasteful to birds. These mimics gain immunity from predation by sailing under false colors. However, immunity is not gained without a price to the model, the mimic, or both. If a bird happens first to catch and eat an unprotected mimic, it may devour several more before an unpleasant encounter with the model teaches it to avoid both the model and its double. The individual bird who seizes the model before it has tasted its mimic may never molest the latter. From this it appears that it is to the advantage of the mimic to remain less abundant than its model. The deceptive resemblance of an unprotected to a protected organism is called Batesian mimicry, for Walter Bates, who discovered this relationship in Amazonian Brazil. Something very similar to this occurs within a species, such as the Monarch butterfly, in which some individuals gain immunity by eating a milkweed that contains a poison, while other individuals, unprotected because they were nourished by innocuous milkweeds, benefit by being indistinguishable in appearance from the former. This situation has been designated "automimicry."

Even when a bird seizes a distasteful insect in its bill only to drop it promptly, the insect may be mortally injured. Except in the cases where an individual bird learns by watching another to avoid a certain protected insect, it appears that every unpalatable species must sacrifice at least one of its members to teach every potential predator to refrain from touching this species. For a rare butterfly, the price of giving all these lessons may be heavy. The more abundant the species, the better it can bear the burden. When two or more species of protected insects resemble one another so closely that the predator who has been punished by one of them will henceforth avoid both or all of them, the costs of educating predators will be more widely spread. This advantage has given rise to Müllerian mimicry, for Fritz Müller, who first recognized it in southern Brazil. Whereas in Batesian mimicry an unprotected species imitates a protected species, in Müllerian mimicry two or more protected species converge in appearance. Mimicry is, of course, not deliberate or conscious but a result of natural selection of random mutations, which favors individuals that most closely resemble a protected species.

Since protected butterflies and their mimics are recognized by their appearance, they should be regarded as warningly colored, although not so obviously as the caterpillars, true bugs, and other creatures with bold patterns or startling hues. Often their patterns are intricate rather than simple and sharp, and birds must learn to avoid them. In contrast to concealing coloration, which benefits the cryptic creatures but not the predators who might be nourished by them, warning coloration benefits both prey and predator: most obviously the former, by saving its life, but also the predator, who is spared the unpleasant or painful consequences of seizing the protected creature by mistake. Moreover, after an initial lesson, a predator does not waste time and energy by pursuing insects that it would not eat. Aposematic coloration might be compared to warning signs posted conspicuously for the protection of the public. It promotes the peaceful coexistence of diverse creatures. Among protected butterflies are some of the most beautiful, and mimics that closely resemble them can hardly be less lovely. These butterflies, and warningly colored creatures of other kinds, are our first examples of how the mutually beneficial relations of organisms promote beauty.

Markings that more or less closely resemble the eyes of vertebrates are frequent on insects, especially butterflies and moths. Although often decorative, they are not without utility. One morning, standing in bright sunshine in a grove of bananas beside rain forest, I watched an owl butterfly (*Caligo* sp.) alight on the big red inflorescence bud of a banana plant. Resting there with head upward, the butterfly pushed its slender proboscis into the white staminate flowers clustered beneath the up-turned bract. While sucking the nectar, it kept its wide wings folded together above its back, displaying the finely vermiculated undersides, marked by a big, black, yellow-rimmed eyespot on each hindwing and a much smaller eyespot on each forewing. Only rarely did the butterfly partly open its wings, giving me instantaneous glimpses of the rich pur-ple and yellow of their upper sides.

While the butterfly was present, the little, stingless, pollen-gathering meliponine bees often flew near but rarely alighted on the flowers. Twice a Long-tailed Hermit hummingbird (*Phaethornis superciliosus*), who was making the rounds of the banana flowers, approached the inflo-rescence where the owl butterfly was enjoying a long feast of nectar, but before it touched a flower a slight movement of the insect's wings sent it off, possibly intimidated by those staring eyespots. After a brief absence, the brown hummingbird returned to the grove, visited a number of other inflorescences, then flew toward the butterfly from directly behind, in-stead of from the side, as on earlier occasions. Unable to see the eyespots because the edges of the wings were now turned toward it, the hermit approached more confidently and supplanted the butterfly without touch-ing it. Possibly the *Caligo* was ready to leave after imbibing nectar for a full half-hour.

Both the Peacock Butterfly (*Inachis io*) and the Eyed Hawk Moth, two European species, bear large eyespots on the upper sides of their hind-wings. The Peacock rests with its wings in contact above its back; the hawk moth with its forewings spread backward to cover the hindwings. Both hide their eyespots while they repose undisturbed; both suddenly expose them when touched or attacked. As told in *Curious Naturalists,* Niko Tinbergen and his coworkers studied the effects of these eyespots on Chaffinches (*Fringilla coelebs*), Jays (*Garrulus glandarius*), Great Tits (*Parus major*), and Yellowhammers (*Emberiza citrinella*). Nearly always, a bird without previous experience with these insects was greatly alarmed when, upon being pecked, one flapped or spread its wings, suddenly revealing eyespots. After giving an Eyed Hawk Moth an exploratory peck, a Chaffinch "jumped back as if stung" when the moth began to display. Lightly pecked by a Jay, a Peacock flapped its wings, so scaring the bird that it jumped straight up into the air and hit the roof of its cage. Nevertheless, after a few minutes it returned and ate the moth. Al-though, after their first surprise, some birds ignored the eyespots and

Owl Butterfly, *Caligo* sp.

ate Peacocks, others developed such an aversion that they would not touch these butterflies, even after the scales bearing the spots had been brushed off their wings.

A series of experiments, made by David Blest and reported by Tinbergen, showed that the more realistic the eyespot, the more effectively it

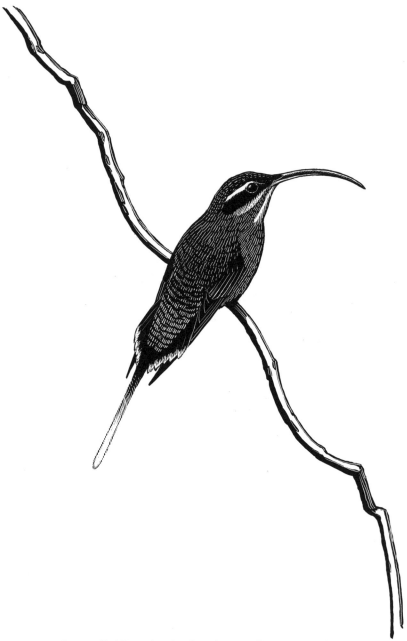

Long-tailed Hermit, *Phaethornis superciliosus*. Sexes alike.
Southern Mexico to Bolivia and central Brazil.

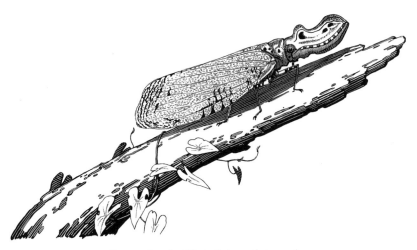

Peanut-headed Bug, *Fulgora laternaria*.

held small birds aloof. Even a simple ring placed on either side of a mealworm made birds hesitate or desist from taking this food; but a model representing a shiny focusing eye was much more frightening. These tests showed convincingly that natural selection could promote the evolution of ever more perfect eyespots that gave more adequate protection to their bearers.

A very different insect equipped with eyespots is the Peanut-headed Bug (*Fulgora laternaria*), widespread in tropical America. The head of this large homopteran extends forward in a projection that vaguely resembles a peanut, or perhaps an alligator's head. The adult insect rests quietly on the bark of a tree, where its pale wings, mottled with gray, make it difficult to detect. If disturbed, it may suddenly fly, emitting a fetid odor, or it may rattle its head against the bark. If roughly treated or knocked to the ground, it spreads its wings widely, revealing on each hindwing a large eyespot not unlike that of an owl butterfly, and doubtless equally intimidating to small birds.

How can we explain the fact that animals which consistently try to remain hidden, or to repel others by their aspect, so often appear attractive or even beautiful to us, which is just the opposite of what we might expect? Should we not more often find them severely plain or repulsive, as some of them are? However, the elements of beauty — form, color, pattern, and texture — are not absent from creatures that try to hide, to warn, or to repel. Feathers have such a pleasing texture that birds are seldom ugly. We enjoy the bright colors that so often proclaim unpalatability or venom. The pigments that color the exposed surfaces of animals tend to be deposited in definite patterns rather than at random. Fear need not diminish our enjoyment of cryptically or aposematically colored organisms, for most are not harmful; they ask only to be permitted to live in peace. When we add to all this our pleasure in using our eyes, seeing colors, recognizing forms and patterns, it is not difficult to understand why creatures that shun observation, or warn that they should not be molested, so often attract us by their beauty.

References

Belt 1888; Brower 1969; Rothschild and Lane 1960; Tinbergen 1958; Urquhart 1960; Wallace 1871; 1872; Weed 1923.

4.
Beauty That Entices

Of all things that are, or are intended to be, beautiful, flowers are the most universally admired. The painting, sculpture, architecture, or music that delights one person may bore or disgust another; an animal that appears beautiful to you may be repugnant to me; but one must be aesthetically inert not to enjoy flowers. Yet flowers are, aesthetically if not biologically, among the simplest of beautiful objects. Their pleasing qualities are form, color, and texture. Their uncomplicated outlines, with radial or bilateral symmetry, rarely confuse vision. Only exceptionally are they unsymmetrical. Their colors, from white to violet in a spectrum of many lovely shades, are enhanced by the satiny or velvety texture of the petals or other parts that bear them. The surfaces of floral parts are frequently unpatterned, or at most marked with simple streaks or spots, or a change of color or its intensity from base to apex. Compared with the complex color patterns of many butterflies, birds, or products of human art, those of flowers are extremely plain. It is just this simple beauty that makes flowers so widely appealing. They offer a lesson in the value of simplicity. And when we recognize, as will become evident in the present chapter, that flowers owe their loveliness to the mutually beneficial interactions of diverse organisms, they teach us another lesson no less valuable.

The essential parts of flowers are pistils and stamens. A pistil consists of an ovary that encloses one or more ovules that develop into seeds, and one or several styles, each bearing a stigma that receives pollen. A stamen consists of a filament, often greatly reduced, supporting the anther that contains the pollen grains. Pistils and stamens may be in the same or separate flowers; in the latter case, the male and female flowers may be on the same or different plants. The simplest flowers, which are mostly wind-pollinated, may consist of little more than these essential organs. Flowers that attract pollinating animals usually have a well-developed corolla, consisting of petals that may be separate or more or less united into a rotate, tubular, or two-lipped structure. Subtending the corolla is a calyx, formed of separate or united sepals which are usually green but sometimes as brightly colored as the petals. Sometimes, as in the poinsettia, bougainvillea, and flowering dogwood, small, inconspicuous flowers are surrounded by large white or highly colored bracts, which substitute for petals in attracting pollinators. Although morphologically distinct from the flowers, functionally and aesthetically they belong to them.

A flower may be structurally simple and readily pollinated by a diversity of visitors, or complex and dependent upon some specialized pollinator. Simple flowers commonly last only a day or two, many of them only a few hours, during which they are pollinated, with the result that a high proportion of them usually set seed. More complex flowers, of which orchids and milkweeds (Asclepiadaceae) are good examples, often remain fresh for weeks or, in certain orchids, even months, awaiting a visit from a suitable pollinator. Even in their natural habitats, many or most of these complex flowers may fail to be pollinated. The few that receive pollen, packed into compact masses called pollinia, compensate for all their barren neighbors by forming pods with many seeds — thousands if not millions of minute seeds in many orchids. For practical no less than aesthetic ends, simplicity has many advantages.

Contributing largely to the beauty of flowers is their richness or depth of color, which in all its delightful variety is produced by only a few fundamentally different pigments. Yellow is contributed by carotene and related lipochromes, dispersed in the floral cells in the form of granules, short rods, or, rarely, long threads. Other colors, from violet and blue to orange, pink, red, and sometimes even yellow, are due to anthocyan dissolved in the cell sap rather than concentrated in solid plastids. The color of anthocyan depends upon the acidity of the medium; red in an acid solution, it becomes blue in a basic medium, as one can readily demonstrate by transferring a fragment of petal from one to the other. Sometimes anthocyan and carotene pigments are present in the same cells, producing a delightful shade of orange. Perhaps more often, they are situated in different regions of a corolla, which in consequence displays two, or rarely more, colors. The flower may be blue with a yellow "eye," or yellow with a red throat.

Since the pigments of petals are within the cells rather than on the surface, to produce rich floral colors light must penetrate to them; light reflected from the surface yields only a silvery glare, as on leaves with thick, waxy cuticles in sunshine. To increase the penetration of light, the epidermal cells of the petal's upper surface, covered with a thin cuticle that retards desiccation, often bulge outward in the form of domes, mounds, or papillae. Thus they serve as light traps; a ray falling on the side of one of these projections at grazing incidence, where reflection is greatest, may glance to an adjacent cell and strike it at an angle more favorable for penetration. Another feature that reduces the surface reflection that masks true colors is the striation of the cuticle by minute ridges, which make the petal's surface optically rough and apparently increase the penetration of light. By such contrivances that are revealed only by a microscope, many flowers achieve a deep, rich coloration and velvety texture that increase their appeal.

Flowers, like people who enjoy them, appeared late in the history of life, but flowers long antedated humankind. If we could have strolled through forests of the early Carboniferous period, over 350 million years ago but 2 or 3 billion years after the first appearance of simple plants, we would have been impressed by the lepidodendrons and sigillarias, long-vanished lycopods that towered high above seed ferns and primitive true ferns of more familiar aspect than the trees. We would have found no flowers, and probably little color other than greens and browns, amid all this exuberant vegetation. Angiosperms, or flowering plants, do not appear in the geological record until millions of years later, in the early Cretaceous period, about 130 million years ago. The earliest of these flowering plants were probably trees which, like the conifers that preceded them and many modern trees of northern lands, were wind-pollinated. Only later in the Cretaceous did flowering plants and animals, mainly insects, discover that they could cooperate to their mutual advantage, a development that led to a vast diversification of both floras and their pollinators, and to the evolution of the kinds of flowers that attract us by their beauty — flowers to which not only botanists but children would apply this name. From the first, the evolution of flowers and their animal pollinators has been closely linked; they may be said to have coevolved.

Pollination by animals brought great advantages. Wind pollination is feasible in northern woodlands composed of a few kinds of trees or often of a single species in almost pure stands. It also suffices for grasses and sedges that grow thickly in open fields and marshes. It would never do in tropical rain forests where scores or hundreds of arboreal species are mixed together, and a tree of a certain kind may be separated from its nearest neighbor of the same kind by many different species. Rain-forest breezes, usually light except in rainstorms that would wash pollen grains to the ground, would waft little pollen between trees separated by many others in full foliage. The great diversity of tropical forests depends strictly upon pollinators that can fly from flowering tree to flowering tree of the same species, past many intervening trees of different species. Moreover, pollination by animals is usually more economical than pollination by wind or water. Great quantities of the pollen so freely shed by conifers and other wind-pollinated trees are finally dropped into rivers, lakes, or oceans many miles from land; only a minute fraction of the pollen grains fertilize ovules. Although animal pollinators must be paid, with nectar, pollen, or both, a much greater proportion of the pollen reaches its proper destination.

Beetles appear in the geological record in the Permian period, about 280 million years ago. They may well have been the first pollinators of flowering plants, which do not enter the record until many millions of years later. Today they play a minor role in pollinating flowers, chiefly such as lack delicate petals and bright colors. Among beetle-pollinated plants are certain aroids, including *Arum, Dieffenbachia,* and *Philodendron,* whose small, petalless florets are crowded upon a fleshy axis, called a spadix, which is enclosed in a thick, convolute spathe. The pistillate flowers are at the base of the spadix, the staminate flowers above them. When the former are ready for pollination, the inflorescence produces heat which, together with odor, attracts small beetles that enter the lower section, where they are held captive until, on the following day, the staminate flowers shed their pollen and the prisoners are released, to escape by crawling over them and becoming covered with grains that they may carry to other inflorescences. Small beetles also come in numbers to certain palms, such as the Pejibaye (*Bactris gasaepes*) and related species, whose massed staminate flowers become much warmer than the surrounding air while they shed their pollen. Even on warm tropical evenings, heat appears to attract these insects strongly. Other flowers pollinated by beetles include the magnolias, sweet-scented shrubs (*Calycanthus*), the great pond lily *Victoria regia,* California Poppy (*Eschscholzia californica*), wild rose, elder, spiraea, Flowering Dogwood (*Cornus florida*), and a few species in the parsley and composite families. The contributions of beetles to floral beauty are not negligible.

Bees, in their great diversity, are the principal pollinators of flowers. Typical bee flowers have spreading petals on which the insects can alight. When the corolla is zygomorphic, or bilaterally symmetrical, the lower lip is often expanded to provide a landing platform. If the dorsal petal is broadest, like the labellum, or "lip," of many orchids, or the standard of certain pealike blossoms, the flower may rotate, by twisting its stalk or inferior ovary, through about 180 degrees, to bring the broad petal below, for the convenience of bees. In addition to many orchids, resupination, as this reversal of orientation is called, is found in *Centro-saema* and *Clitoria* of the Leguminosae.

Bees recognize their flowers by both color and scent. Perhaps most often color draws their attention at a distance, and after flying closer they recognize the species by its odor. However, when flowers are screened by foliage or other obstructions, while scents are widely diffused, the latter may be the primary attraction. The colors and scents that draw bees to blossoms are as pleasing to us as they are attractive to them. Nevertheless, their color vision differs from ours; red is not distinguished by them, but they are sensitive to yellow, blue-green, blue, and violet, as well as ultraviolet that we cannot see. Accordingly, bee flowers are rarely pure red; those which approach this color reflect a mixture of shorter waves that make them purplish or mauve. Bee flowers often display "honey guides" in the form of stripes converging on the spot where the insect must insert its proboscis to suck up the nectar; or a contrasting color or "eye," often a patch of yellow or white at the center of a blue or lavender corolla, or a reddish throat in a yellow corolla. Flowers that appear uniformly colored to us may present to bees a pattern that can be revealed to human eyes by photographing them in ultraviolet light.

Experiments reported by Niko Tinbergen in *Curious Naturalists* demonstrated that bees prefer flowers with honey guides. They tend to alight on the corolla's margin, then follow the guides to their reward. Bees are efficient pollinators because they tend to concentrate on a single species while it blooms freely, thereby transporting the pollen that adheres to their often furry bodies to flowers that will be fecundated by it. The honeybee that finds a rich source of nectar returns to her hive and, by means of a display described by Karl von Frisch in *The Dancing Bees,* reveals to her sister workers the direction and distance they must fly to reach the same flowers, whose kind is revealed by the scent carried on her body. Bumblebees and stingless meliponine bees do not convey information to other individuals in this manner. Although some bees are attractively marked, often with yellow and black, they are not the most brilliant of insects. Exceptional are the orchid-pollinating euglossine bees, shining in iridescent green or blue. Species of plants pollinated by bees at low altitudes often depend upon butterflies to pollinate their flowers on cold alpine heights.

In contrast to bees, wide-winged butterflies are frequently more beautiful than the flowers they visit. Like bees, they are attracted to flowers by both scent and color, which may be red or orange as well as yellow or blue, as butterflies have a visual spectrum that includes all the colors we can see, plus ultraviolet. Butterfly flowers open by day, while those pollinated by their relatives, the mostly nocturnal moths, expand in the late afternoon or evening and are often white. Frequently they diffuse a heavy fragrance, which the moths, extremely sensitive to scents, follow upwind. Hawk moths are active in the evening twilight, when while hovering they may be confused with hummingbirds, which continue to sip nectar after most diurnal birds have retired to their roosts or nests. Lepidoptera carry, coiled up beneath their heads, the probosces through which they suck nectar. Those of moths are often extremely long. The spurs of the orchid *Angraecum sesquipedale* of Madagascar are ten to fourteen inches (25 to 36 centimeters) in length. Writing of these flowers more than a century ago, Alfred Russel Wallace predicted, correctly, that a moth with a proboscis of corresponding length would be found on that island.

Long-tongued flies of the families Bombyliidae and Syrphidae visit the same types of flowers that bees frequent, attracted, like bees and butterflies, by color and fragrance. Syrphids are so frequently seen taking nectar and pollen that they are called "flower flies." To what degree these flies have promoted the evolution of the features that make flowers delightful to us, we do not know. Probably bees, the principal pollinators, are largely responsible for the presence of such flowers, and the dipterous flies subsequently learned to profit by them. The short-tongued flies of many families that visit certain flowers are lured to them by their frequently putrid odors, as of carrion, dung, and other smells disgusting to us. The colors of these flowers are almost equally unattractive, often dark red or purplish brown, suggestive of decaying flesh; at best they are greenish yellow or white.

Chief among the avian visitors to flowers are hummingbirds, sunbirds, honeyeaters, sugarbirds, Bananaquits, honeycreepers, lories, and lorikeets. Most specialized for flower visitation and most studied are the hummingbirds, of which about 315 species inhabit the continents and major islands of the Western Hemisphere, mostly within the tropics, with a few migratory species reaching high northern and southern latitudes. Typical hummingbirds are small to very small, clad in brilliant metallic plumage, with long, slender bills and long, protrusile white tongues for sucking nectar. Typical hummingbird flowers have long, tubular corollas, usually without the expanded landing platform of bee flowers, and they lack fragrance. As a rule, they are freely exposed in an erect or hanging inflorescence, with no convenient perches where the birds can rest while they drink nectar. This forces them to hover on wings beaten into a haze, as no other bird can. Such rapid exercise is expensive of energy; when occasionally they find a suitably situated perch, hummingbirds may rest upon it while they drink. Sicklebills (*Eutoxeres* spp.) have, for hummingbirds, exceptionally strong feet and regularly perch while visiting flowers of species of *Heliconia,* on which they specialize.

Why have hummingbirds such long bills, such great ability to hover motionless, such extraordinary maneuverability in the air? The origin of the hummingbird family is obscure. They have long been classified in the same order, the Apodiformes, as the swifts, which also have exceptional powers of flight. However, hummingbirds and swifts use these powers in wholly different ways, and no two families could differ more strikingly in appearance. Some hummingbird flowers evolved from bee flowers. In several ways, hummingbirds are more efficient pollinators than bees: they do not carry lumps of pollen to their nests, as bees do to their hives, where it fails to benefit flowers; they do not eat pollen; they can carry larger loads of pollen on their bigger bodies, from which it may rub off on the stigmas of other flowers; they fly more rapidly and for longer distances than bees, thereby spreading the pollen through a wider population of plants, with the genetic benefits of outcrossing; they are more dependable than bees, because they are active throughout the year, in all kinds of weather, even at high altitudes where low temperatures immobilize bees.

Accordingly, it was advantageous to many kinds of plants to discourage the visits of bees to their flowers while favoring those of the ancestral hummingbirds, which they did by evolving long, narrow floral tubes without landing platforms. Simultaneously, hummingbirds, which appear to be closely related to passerines, developed longer bills and improved their ability to hover. They became adept in flying backward, to extract their bills from long tubes, and in maneuvering sideward and turning abruptly as they shifted from flower to flower in the same or different inflorescences. A hummingbird's visit to a small flower often lasts but an instant; by such aerobatics it greatly increases the number of flowers it visits in a short interval. It could not gather nectar so rapidly if it perched for each flower.

Much has been written, and much experimental work has been done, on hummingbirds' preference for colors. Hummingbirds, whose visual spectrum is much like ours, visit flowers, or artificial feeders, of all colors, including white. They are much more interested in the amount or quality of the nectar or sugar-water offered to them than in the color of the container. Active and curious, they readily discover new sources of sweetness, often on introduced plants with floral structures very different from those of New World hummingbird flowers. The predominance of red in the hummingbird flowers of temperate North America is related to the migratory or nomadic habits of the relatively few members of the family that breed there. It is to the birds' advantage to have all or most of their flowers of the same color, so that they may promptly recognize them wherever on their travels they may be, just as it is advantageous to motorists to have uniform traffic signals everywhere. And red is the color that most attracts attention amid green foliage. In the tropics, where hummingbirds are permanently resident (although they may wander rather widely as now here, now there, flowers bloom) hummingbird flowers are often red but not so predominantly as in the north.

In the tropics and subtropics of the Old World, from Africa, where they are most numerous in species, through the Middle East, India, and Indonesia to northern Australia, 116 species of sunbirds are the ecological equivalent of the New World's hummingbirds. Both families are well represented from warm lowlands to the greatest heights where flowers bloom. Like hummingbirds, sunbirds are mostly small and attired in colorful, glittering metallic plumage, especially the males. Their bills are long, slender, and downcurved; their tongues are mostly tubular with divided tips. Their nostrils are covered with flaps, or opercula, to keep out pollen grains. Their sharp-clawed feet are stronger than those of hummingbirds, for instead of hovering in the air they usually cling, often in what appears to be a strained posture, while they extract from flowers the nectar on which they largely subsist, and at the same time pollinate the flowers, often the red or orange blossoms of trees of the genera *Erythrina, Spathodea,* and *Symphonia.* Sunbirds supplement their diet of nectar by gleaning small insects from foliage and inflorescences. Unlike hummingbirds, to whose breeding habits we shall give attention in Chapter 7, sunbirds form monogamous pairs. The female builds a hanging or embedded, purse-shaped nest with a side entrance, and incubates the two or three eggs without the male's help. He defends the flowers that support her and aids in feeding the nestlings.

Important pollinators throughout Australia and in New Guinea, islands of the southwestern Pacific, New Zealand, and Hawaii (where 3 or 4 of the 5 species formerly present have become extinct) are many of the 169 species of honeyeaters of the family Meliphagidae. Averaging much larger than hummingbirds and sunbirds, they are much more diverse. Only a few can compare in brilliance of coloration with their African

Nectar-drinking birds that pollinate flowers. Top left: Gould's Sunbird, *Aethopyga gouldiae,* male, India to Indochina. Top right: Booted Racquet-tail, *Ocreatus underwoodii,* male, Venezuela to Bolivia. Center: Regent Honeyeater, *Xanthomyza phrygia,* sexes similar, southeastern Australia. Bottom: Varied Lorikeet, *Psitteuteles versicolor,* sexes similar, northern Australia.

and American counterparts; many are dull green, gray, or brown, which on some species is relieved by patches of black, white, or yellow. Most have fairly long, decurved, sharp bills and protrusile tongues with brush-like tips that serve well to extract nectar from flowers, which liberally dust their heads and beaks with pollen. Like the hummingbird flowers of America, Australian flowers specialized for pollination by honey-eaters have long, narrow, tubular corollas that are often red or yellow. Trees of the families Proteaceae and Ericaceae are largely dependent upon honeyeaters for pollen transfer. Like other nectarivorous birds, honeyeaters enrich their diet with insects, and some eat much fruit, which hummingbirds and sunbirds rarely or never do. As a family, honey-eaters are less closely bound to flowers than these smaller birds.

In Australia, New Guinea, and neighboring islands, fifty-five species of lories and lorikeets share with honeyeaters the work of pollinating flowers. Clad in green, red, yellow, blue, and purple in varied patterns, these small parrots are among the most ornate members of their family. Their brush-tipped tongues have long been regarded as organs for extracting nectar, but their chief function appears to be gathering pollen and pressing it into pellets or cakes that can be readily swallowed. When nectar is abundant, they take that, too, and grow fat upon it. They also eat fruits, and some species include seeds, buds, larvae, and mature insects in their diets. Unlike other nectar-drinking birds, which tend to forage alone or in pairs, lories and lorikeets fly from flowering tree to flowering tree in noisy, chattering flocks of sometimes hundreds of brilliant birds.

These four are the most important families of avian pollinators, but scattered over Earth are many other birds that may fecundate flowers while they seek nectar. Widespread over the warmer parts of the American continents and the Antilles is the Bananaquit (*Coereba flaveola*), which is now sometimes included in the wood warbler family (Parulidae), but is perhaps best kept in a family all its own. This diminutive bird, with dark upperparts, yellow underparts, and a prominent white eyebrow, subsists chiefly upon nectar that with a sharp, decurved bill it extracts from flowers of many kinds, and tiny insects and spiders that it gleans from foliage and blossoms.

Formerly placed with the Bananaquit in the family Coerebidae but now classified in the tanager family, fifteen species of lovely little honey-creepers range over tropical America, chiefly at low and middle altitudes. With forked, fringed, almost tubular tongues and sharp-pointed bills that vary from rather short to fairly long and downcurved, they probe many flowers while clinging beside them instead of hovering. However, fruits, fruit juices, and insects are the mainstay of most honeycreepers, as of other tanagers. Among wood warblers, the Tennessee (*Vermivora pere-grina*), at least in its winter home in tropical America, is a persistent nectar-sipper and possibly helps to a limited degree to pollinate flowers. Numerous other birds that relish nectar probably more often harm than help the flowers, as will be told beyond.

In southern Africa, two species of sugarbirds (*Promerops* spp.), belonging to a family all their own, are important pollinators. Dull brownish, long-tailed birds with long, decurved bills, they probe flowers of the numerous trees of the genus *Protea*. In addition to nectar, they take many insects, especially while the female is forming her eggs; and both parents feed the two nestlings in open cups, built by her alone.

Nectarivorous birds are frequently reluctant to share their flowers. Hummingbirds, sunbirds, and honeyeaters have a reputation for pugnacity; even lories, who forage in larger flocks, are said to be quarrelsome. Some honeyeaters spend so much time driving other birds away that their nests are neglected. Tennessee Warblers become more aggressive at flowering trees than when they search for insects. Pairs of Common Amakihis (*Hemignathus virens*, a Hawaiian creeper) subdivide their territories while foraging and, especially while nesting, the female actively defends her area against her mate. One may watch many birds of different species eating together in a berry-laden tree with scarcely any friction between them, but a flowering tree often presents a far less orderly scene. Nectar does not sweeten the tempers of birds who imbibe it freely.

Why are nectar-drinkers so much less pacific than frugivores? One reason may be that berries tend to be more abundant, or more readily gathered, than nectar. Usually a bird can tell at a glance, by the color of a berry, that it is ripe and ready to be eaten. A nectar-drinker may have to probe a flower to learn whether it contains a drink or has recently been drained, and this takes time and energy. If it can hold competitors aloof, it is more likely to find the cup full. However, a profusely flowering tree may draw too many visitors for one bird to expel, and perforce it shares the bounty with others.

Among mammals, bats are the main pollinators. A slender, elongate muzzle, long protrusile tongue with a papillate surface or brushlike tip, and absence or reduction of front teeth facilitate visits to flowers by tropical long-tongued fruit bats (Macroglossinae) and long-nosed bats (Glossophaginae), which derive most or all of their nourishment from this source. Bat flowers give free access to these flying mammals by their exposed positions, hanging in the air on long branches or ropelike peduncles, growing on bare trunks or limbs, or standing free at treetops.

These flowers are large, nocturnal, often dingy white or greenish in color, with wide mouths that make generous amounts of pollen or nectar readily available to the animals, who are attracted by odors often disagreeable to humans. Among bat-pollinated flowers are those of the Kapok or Silk-cotton Tree (*Ceiba pentandra*), Calabash (*Crescentia cujete*), Sausage Tree (*Kigelia pinnata*), Candle Tree (*Parmentiera cerifera*), and many others. Mice and small marsupials pollinate a few night-flowering plants. Mammals that at night are attracted to flowers by scents often disagreeable to us rather than by colors have contributed little to floral beauty and need not detain us longer.

Colors and scents draw to flowers pollinators who require more substantial rewards for their services. Most often they receive nectar, pollen, or both. In addition to the sugars that make nectar sweet and are its main attraction to animals of many kinds, it often contains, more obscurely, a number of ingredients that nourish those who drink it, including amino acids, proteins, lipids or fats, and vitamins — substances that help to make honey a healthful food. Proteins and lipids are rather rare in nectar, amino acids more frequent. If not all the essential kinds are present in a single species of flower, by visiting different species an insect can collect a full complement of these indispensable builders of proteins. Birds, who have sources of nourishment other than flowers, depend less than bees and butterflies on nectar for their amino acids. In addition to nutritive substances, alkaloids are present in detectable amounts in the nectar of flowers pollinated by bees but appear to be absent from those frequented by butterflies and moths, which are less resistant to these noxious chemicals. The alkaloids in bee-flowers may deter the visits of the less dependable lepidopteran pollinators.

Although in the absence of nectar, hummingbirds and most other birds, with the notable exception of lories, have little incentive to visit flowers, many kinds offer only pollen to bees. The grains may be freely shed and readily accessible from anthers of many types, or they may escape only through apical pores that make them more difficult to procure. Among plants devoid of nectar that guard their pollen well in anthers that open only by narrow pores are many species in the melastome and nightshade families, as well as golden-shower trees of the genus *Cassia* in the pea family. To extract pollen through pores, bumblebees and other fairly large kinds seize an anther in their mandibles, or press their closed mandibles against it and, emitting a high-pitched buzz, vibrate it so rapidly that a puff of pollen escapes, dusting the bee's body. Often the vibration is so strong that it can be felt by holding the twig that bears the flower. If the anther is long and slender, a bumblebee, while vibrating, pulls outward as though massaging or milking it, a treatment that leaves brown scars. Smaller bees, unable to shake massive anthers, glean grains left exposed after the departure of the big ones.

However the pollen is procured, bees of many kinds amass it in the pollen baskets on their hindlegs, then fly to their hives with prominently bulging lumps of whitish or golden grains protruding outward, one on each side. Mixed with nectar, often from different flowers, the pollen nourishes the larval bees in their brood cells. All this pollen is lost to the plants, which must produce enough of it to fertilize their flowers as well as to satisfy their pollinators. Grains that adhere to the bees' bodies in places from which they are not readily shifted to the baskets are most likely to be rubbed upon the flowers' stigmas. Some flowers solve the problem by producing two kinds of pollen in anthers that differ in shape and situation. Those offered to the bees as food are shaken upon the insects' ventral surfaces, where they are most readily moved to the baskets, while those for pollination fall upon the bees' backs. Among plants with two kinds of pollen are species of *Melastoma,* whose pollen from the food anthers is said to be incapable of germination; and species of *Cassia,* in which both kinds of pollen germinate in sugar solutions and are capable of fertilizing the ovules.

Unable to eat solid food, and neglectful of their progeny, most butterflies visit flowers only for their nectar. Exceptional among lepidoptera are tropical butterflies of the genus *Heliconius,* which collect pollen on the proboscis and mix it with nectar, thereby releasing amino acids which the insects absorb without having to chew or digest. With this source of nourishment while in the adult stage, heliconians are exceptionally long-lived butterflies, often surviving for six months. Nourished by pollen, they can produce five times as many eggs as when they are restricted to a diet of nectar alone.

Different pollinators make different demands upon the plants they serve. From the time they hatch until they die, many bees are wholly dependent upon nectar and pollen from flowers for nourishment. Butterflies are less closely bound to flowers; as adults they may suck fruit juices and even liquids from decaying matter; as caterpillars they gnaw leaves and other vegetable tissues, exceptionally those of flowers. Probably, to be perfectly fair, after they acquire wings they should pay for their larval food by pollinating the species of plants whose foliage they disfigure or destroy; but nature is not always just. Thus, heliconians that as caterpillars feed upon passionflowers as adults pollinate the vines *Anguria* and *Gurania* in the gourd family (*Cucurbitaceae*), neglecting the vines that nourished their immaturity. Of all the major pollinators, birds demand least of flowers. With more varied diets that include many insects and often also fruits, they ask of flowers only sugary nectar, which need not contain accessory nutrients such as amino acids and vitamins. For economy, as well as for the wide dispersal of their pollen, plants should favor hummingbirds as their pollinators.

Although nectar and pollen are the flowers' usual rewards to their pollinators, a few attract them with other gifts. Species of *Clusia,* epiphytic shrubs and small trees of the family Guttiferae, widespread in tropical American woodlands, offer to bees, for the construction of their nests, resins secreted by their flowers. How the workers avoid becoming hopelessly ensnared by this extremely adhesive stuff, like Br'er Rabbit on the Tarbaby, I fail to understand. It is almost impossible to remove from wood or metal, and when, in a battle with ants that attempt to invade their hives, the bees fasten tiny pellets of it upon their enemies, the latter are permanently disabled. Other plants that use this rare enticement to pollinators are vines of the genus *Dalechampia* of the euphorbia family, widespread in tropical America. The viscid resin is secreted by a glandlike structure above the staminate flowers of the compact inflorescence, surrounded by conspicuous bracts. A Brazilian orchid, *Maxillaria divaricata,* secretes on its labellum generous amounts of wax attractive to insect visitors.

Shiny male golden, or euglossine, bees fly to orchids and certain other plants to collect components of the scents that they use for courtship displays. Alighting upon the flowers, the bees brush the surface with special pads on their front feet, then hover while they transfer what they have collected to cavities in their swollen hindlegs. They may repeat this process for hours. Each species of bee has a special fragrance that it compounds with ingredients collected from several kinds of flowers. It employs this scent not, as one might suppose, to entice partners of the opposite sex, but to lure other golden males, who dart about in a small swarm that attracts females to this gathering reminiscent of the courtship assemblies, or leks, of certain birds.

Although most visitors to flowers earn the nectar, pollen, or other things that they take by transferring pollen from bloom to bloom, not a few carry off the plants' bounty without paying for it. By piercing the base of a corolla, insects and birds frequently reach nectar without touching anthers or stigmas, then fly away with their stolen nourishment. Bees frequently perforate flowers which have tubes so long that they cannot reach the nectar in the "legitimate" manner. Some birds live by violating the mutually beneficial relations between plants and their pollinators. Chief among them are thirteen species of flower-piercers of the genus *Diglossa,* long included in the honeycreeper family (Coerebidae) but recently placed with the emberizine finches. These little birds, clad mostly in subdued blue or black, are widespread in the highlands of tropical America, where they are often found on flowery mountaintops and woodland openings. Seizing the base of a corolla in its short, uptilted bill, the flower-piercer holds it steady by placing the hooked tip of its upper mandible over it, while its sharp lower mandible pierces the tissue and the fringed tongue extracts nectar through the tiny perforation. A Bananaquit simply pushes the sharp tip of its short, downcurved, hookless bill through the base of a corolla so long that it cannot reach the nectar from the mouth. With a bill shorter and sharper than those of most hummingbirds, the lovely Purple-crowned Fairy (*Heliothryx barroti*) pierces a wide variety of flowers that are sometimes larger than itself. I have not seen it visit any in the legitimate manner. Even longer-billed hummingbirds that usually pollinate their flowers not infrequently perforate those whose size or configuration makes it difficult for them to reach the nectar in the usual way.

Less often than they are defrauded, flowers deceive their pollinators. Orchids, with their immense diversity of floral forms and modes of pollination, include a number of these cheaters. Flowers of the European *Ophrys* and the Australian *Cryptostylis* so closely mimic the females of certain bees and wasps, not only in their forms but likewise in the sexually attractive odors that they diffuse, that the males of these insects attempt coition with them, in the act picking up pollinia that they may carry to the stigmas of other flowers. Lady's slipper orchids (*Cypripedium* spp.) induce small insects to enter their inflated lips, from which they can escape by passing over the column and depositing or picking up pollen, without finding nectar as a reward for their services. Other flowers, including *Parnassia* of the saxifrage family and *Paris* of the lily family, attract pollinating flies with glistening streaks, spots, or corpuscles which yield them no nourishment. Still others, such as Dutchman's pipe (*Aristolochia macrophylla*) and arum, entice flies to enter a trap, where they are detained, fasting, for a day or two until the anthers split and they are released with loads of pollen.

Despite all these artifices, tricks, and deceptions, it remains true that most flowers offer a reward to visitors who at least do not try to carry away pollen or nectar without earning it, and that we owe a larger part of nature's beauty to the prevalence of this exchange of benefits.

At least one animal helps flowers to fulfil their biological function of perpetuating their species for an immaterial reward, enjoyment of their beauty. This is *Homo sapiens,* which sometimes hand-pollinates choice

Birds that "steal" nectar. Top: Purple-crowned Fairy, *Heliothryx barroti,* sexes similar, southeastern Mexico to southwestern Ecuador. Center: Bananaquit, *Coereba flaveola,* sexes alike, southern Mexico and West Indies to northeastern Argentina. Bottom: Indigo Flower-piercer, *Diglossa indigotica,* sexes similar, Colombia and Ecuador.

cultivars but more often leaves insects or birds to accomplish this, while giving the plants more comprehensive care, sowing them, cultivating them, protecting them from harmful insects or fungi, spreading them widely over Earth. By selection, human beings frequently increase the number of a flower's petals at the expense of their reproductive organs, thereby making them bigger and more colorful, but often with loss of the simple symmetry that gives charm to little wildflowers. Many flowers cosseted by humans no longer yield viable seeds and must be propagated by cuttings, bulbs, tubers, or other vegetative parts. These flowers appear to have lost their natural function. Nevertheless, they fulfill this function in a novel way, by becoming so beautiful that they assure the propagation of their species or variety without the need to set seed. On the broader view, the relation between people and ornamental plants has much in common with that between flowers and their natural pollinators. In both cases, each participant in the transaction reaps substantial benefits, for to be gladdened by beauty is no slight reward.

Not only directly, by their presence, do flowers embellish the world of humans, but in large measure also indirectly, as subjects of artistic design. One need only examine the textiles, ceramics, and other artifacts in one's own household to be convinced that floral designs, often combined with foliage or fruits, are used for decoration more frequently than anything else. But how often do we reflect that without the pollinating activities of insects and birds, these pleasing motifs would not be available?

The animals that disperse seeds are nearly always different from those that pollinate flowers; only honeycreepers, honeyeaters, and a few other birds that take both nectar and fruits do both. Just as some flowers trick insects into pollinating them without recompense, so a much larger number of plants surreptitiously exploit animals to disseminate their seeds without payment. These are plants whose usually small, dry pods, capsules, loments, achenes, or other inedible fruits are equipped with prickles, hooks, or sticky secretions that attach them to the fur of animals or the clothes of humans who carelessly brush against these usually annoying propagules. The extremely varied devices by which fruits win free rides are fascinating to study; but such fruits are rarely beautiful or colorful; it is to their advantage to lurk obscurely, like the blood-sucking ticks that give their name to certain of them, so that potential carriers will not notice and avoid them.

Very different are the fruits that reward disseminators, the berries, drupes, pomes, and capsules with arillate seeds. Far from trying to remain unseen, they advertise their presence by colors that contrast with the green foliage amid which they grow. Their colors range from white through red, orange, yellow, pale blue, cobalt, and purple to black. Often they are glossy, or they may be frosted with a delicate, powdery bloom. Although, as a rule, fruiting plants are less spectacular than flowering plants, exceptions occur. Generous clusters of bright red fruits shining amid green leaves, as in the Fire Cherry (*Prunus pennsylvanica*), mountain ash, or rowan (*Sorbus* spp.), and holly (*Ilex* spp.), make a beautiful display. In the rain forest over which I look grow two species of *Cephae-*

lis, in the coffee family, whose small, deep blue berries are displayed between two bright red bracts, from which small birds, especially manakins, pluck them.

Berries and drupes usually have thin skins. Birds peck directly into them, swallow them whole, or pluck them and remove the skin before they ingest them. Arillate seeds are produced in tough or woody capsules which protect them until they ripen and the pod spontaneously opens. Such protection seems necessary because arils, usually rich in oil but poor in sugar and starch, are more eagerly sought by birds than are often watery berries, and if unprotected might be prematurely eaten. The aril itself is a fleshy outgrowth from the base of a seed that partly or completely covers a seed coat which protects the embryo from digestion while in the alimentary tract of a bird. When, at its own good time, the capsule dehisces, it exposes one or many arillate seeds of a color that contrasts with that of the pod. The aril may be white, red, orange, yellow, or black; the pod, red, yellow, green, or brown. The large arillate seeds of trees of the nutmeg family are outstandingly attractive. The two brownish valves of a pod of *Virola* spread apart to reveal a shiny brown seed partly visible between the strips of an irregularly branching red aril. Completely enclosing its seed like a sleeve, the bright red aril of *Composoneura sprucei,* a small tree of the rain forest, hangs like a little red lantern between two separated, light yellow valves. The pale brown seed inside the aril is prettily streaked with deep brown. Pods of epiphytic shrubs and small trees of the genus *Clusia* spread open like a flower with from four to about twelve petals, according to the species. On the inner face of each petal, or valve, is a mass of orange to red arils in which tiny seeds are embedded. Birds are so avid for these seeds that they can hardly wait for the pods to open but with slender bills extract them through the first fissure between separating valves.

Small birds swallow whole small arillate seeds or may peck the arils from larger ones, which bigger birds gulp down entire. After digesting arils or the fruit pulp in which small seeds are embedded, the bird either ejects the resistant seeds through its mouth or voids them in its excreta, often within half an hour of swallowing them. Meanwhile, it may have flown to a distance. If the seeds, still viable thanks to the coats that protected them from the birds' digestive juices, fall in appropriate spots, they may germinate and grow into large trees or epiphytic shrubs. In this manner, birds spread far and wide the plants that offer food to them.

Just as some animals that drink nectar fail to carry pollen, so some birds that exploit fruiting trees fail to disseminate their seeds. Prominent among these seed predators are parrots, who prefer the more nutritious

Color plate 2 (on following page).

Top: four butterflies unpalatable to birds — left to right, *Heliconius narcaea*, *Melinaea ethra*, *Lycorea cleobaea* (Danaidae), *Perrhybris pyrrha*, female. Middle, left to right: Andean Cock-of-the-Rock, *Rupicola peruviana*, male, Andes from Colombia to Bolivia; Guianan Cock-of-the-Rock, *R. rupicola*, male, northeastern South America. Bottom: three toxic frogs — left to right, *Dendrobates pumilo*, Costa Rica; *Phyllobates latinasus*, Colombia; Golden Frog, Panama.

embryos and endosperm of seeds to the soft and often watery pulp of berries and drupes. With thick, powerful bills, hinged above and below as few avian bills are, the larger of them can open tough pods and crush hard seed coats to extract the contents. Seeds attractive to parrots, grosbeaks, and other embryo-eaters are often dispersed by wind, gravity, or small mammals that may drop or bury at a distance from the parent tree those they do not eat. Often the fruits that contain these seeds offer nothing to frugivorous birds, so that, on the whole, seed predators do not compete seriously with true frugivores, at least in tropical America where I have watched them. They are not likely to disrupt the mutually beneficial association of fruiting plants and frugivorous birds.

Fruit-eating birds include a large share of Earth's most beautiful species: birds of paradise, tanagers, honeycreepers, cotingas, manakins, barbets, toucans, New World trogons, fruit doves. The largely or wholly insectivorous families, such as woodcreepers, ovenbirds, antbirds, flycatchers, wrens, and vireos are, on the whole, much more plainly attired. Birds of prey are rarely colorful. Many factors determine the coloration of animals, and one wonders how far fruits are responsible for the brilliance of the birds that consume them abundantly. The brightest colors of feathers, red, orange, and yellow, are due chiefly to carotenoid pigments, while green is the effect of yellow pigment overlying blue, which is produced by the fine structure of the feathers rather than by pigment. Birds, like other animals, are unable to synthesize carotenoids but must derive them from plants or from animals that have eaten plants. The carotenoids that frugivorous birds acquire directly from plants appear to be largely responsible for their brilliant coloration. Another important factor may be the sunlight to which frugivores are exposed while they forage in the crowns of trees or fly from one to another. Birds that pass much time in the forest canopy tend to be much more brilliant than those that lurk in the dim undergrowth. We know that diet influences directly the coloration of flamingos and canaries. The degree to which fruits are responsible for the genetically controlled color patterns of frugivorous birds is a question worthy of investigation.

To stand in morning sunshine before a tropical tree laden with ripe berries or pods exposing arillate seeds is a rare delight. Birds of a score of kinds are continually arriving, eating all they want, and departing. Among them are tanagers, honeycreepers, finches, vireos, thrushes, flycatchers, cotingas, manakins, and woodpeckers, with an occasional motmot, toucan, trogon, or pigeon. What a colorful, ever-shifting display they make in the Sun's bright rays! Each plucks its berries in its own way, perching beside them or seizing them while hovering or darting past. With fruits so plentiful, competition for them is minimal. When occasionally one bird supplants another at a cluster of berries, the individual so displaced flies to a neighboring cluster and continues to eat. No bird tries to exclude another from the tree; fights are extremely rare.

At such a gathering of colorful birds, the thoughtful watcher recognizes beauty of two kinds. In addition to the beautiful plumage of the attendants, one delights in the beautiful relations between all these birds and the tree that offers its bounty to them. This is the way the living world should be; the way it might have become if planned by a benevolent Creator instead of being left to the hazardous course of random mutations and merciless natural selection. Reluctantly, one tears oneself away from this heartening spectacle of life released for a while from conflict. One walks away with the conviction that a large part of nature's beauty springs from reciprocally beneficial interactions between plants and animals — flowers and their pollinators, fruits and seed dispersers, ornamental plants and horticulturists — and that this beauty is spiritual or moral as well as sensuous. Cooperation between animals and plants promotes, in a substantial segment of the living world, the harmony that is the highest goal of moral endeavor.

References

Armbruster and Webster 1979; Campbell and Lack, eds. 1985; Forshaw and Cooper 1977; Gilbert and Raven, eds. 1975; Grant and Grant 1968; Kamil and van Riper 1982; Skutch 1971; 1973; 1980a; 1980b; 1983b; Tinbergen 1958; von Frisch 1954; Wallace 1871; Wickler 1968.

5.
Sexual Selection:
A Preliminary Survey

The preceding chapters considered beauty arising from interactions of unrelated organisms: flowers and pollinators, fruits and seed dispersers, predators and prey. Now we turn to beauty promoted by interactions between members of the same species. This introduces us to the subject of sexual selection, the concept of which we owe to Charles Darwin, whose breadth of vision was balanced by his attention to details. After he presented, in *The Origin of Species,* published in 1859, his theory of evolution by the natural selection of random heritable variations, he recognized that the more widespread modes of selection, which determine the survival and fecundity of organisms, were hardly adequate to account for the weapons and ornamentation of many animals, particularly when they differed conspicuously in the two sexes of the same species. In *The Descent of Man and Selection in Relation to Sex,* published in 1871, he examined in detail the methods and results of sexual selection.

Darwin began his inquiry by distinguishing between primary and secondary sexual characters. The former consist of the reproductive organs themselves and whatever accessory features are indispensable for the insemination of the female by the male. Other structures present in only one sex, such as the mammary glands of mammals, the pouches of marsupials, and perhaps even the pollen baskets on the legs of female bees, might also be considered primary sexual characters, for without them mammals or bees could not rear their offspring, but they grade into the secondary sexual characters. These consist, above all, in sexual differences not essential for the functioning of the reproductive organs, such as the greater size and strength of the male, the weapons with which he confronts his rivals, his bright colors and adornments, his songfulness, and whatever else gives him an advantage in achieving paternity and transmitting his genes to posterity. These features by which males differ conspicuously from females of their species are certainly not indispensable for reproduction, which is effected adequately by many animals of which the male and female hardly differ except in their primary sexual characters; they are significant chiefly in the competition of males (or, in general, the sex that takes the initiative in courtship) for sexual partners, and in the long-term consequences of such rivalry. The theory of sexual selection is concerned with the origin of the structures and behaviors that confer advantage in reproduction or, more briefly, how animals win momentary or more permanent mates.

In Darwin's day, the prolonged, patient study of free animals in their natural habitats that is now frequent had hardly begun. All the more recent observations of the courtship of animals of many kinds in many lands, plus the critical examination of Darwin's theory, notably by Julian Huxley, have led to certain modifications of a basically sound insight. We now distinguish between intrasexual and intersexual selection. Extreme examples of the former are animals of which the males fight for possession of females, or of territories that attract them, as occurs in the more advanced branches of the animal kingdom, from insects to mammals and birds. The purest expression of intersexual selection is the free, uncoerced choice by the female of a partner, as most obviously occurs at courtship assemblies of male birds. These extremes grade into each other; intrasexual and intersexual selection overlap. Harsh fighting may be partly or wholly replaced by nonviolent display between rival males; intrasexual selection occurs even when males vie at a distance from one another for the attention of females, for some will probably be more successful in attracting them, resulting in the differential reproduction which is the true meaning of biological selection. Moreover, a greater or lesser measure of intersexual selection, or female choice, is possible in most, if not all, mating systems, for it would be difficult for a male bird to retain a female who wished to fly away; and even at harems of belligerent male mammals, a determined female might escape.

Paying attention to outstanding features and neglecting minor deviations, we may recognize three basic types of sexual selection (Table 1). The first is intrasexual selection, characterized by the "law of battle." Males not only bluff and display in the face of their rivals but, if more or less equally matched, may attack each other violently, often resulting in injury and even death. The victor in these encounters remains in possession of one or more females, who rather passively accept the outcome of the contest. The victorious male will probably transmit to his progeny the qualities that gave him supremacy. His sons and grandsons

Table 1.　*Types of Sexual Selection*

Type	I Intrasexual	II Unilateral Intersexual[1]	III Mutual Intersexual
Characteristic	The "law of battle"	The contest to charm	The quest of a partner
Quality proximately selected	Belligerence	Conspicuousness and beauty	Compatibility
Activity of males	Fighting and hostile display	Visual and/or auditory display: posturing, dancing, calling	Visual and/or vocal display; often obscure
Activity of females	Mostly passive acceptance	Active choice	Sometimes same as male's
Duration of union	Until after coition	For coition only	Prolonged, often lifelong
Effects on male	Development of weapons, often also of exaggerated size	Development of bright colors and ornaments, vocal and/or instrumental sounds, and/or bizarre antics	Involvement in family care, sometimes increase in brilliance and song
Effects on female	Sometimes acquires male's weapons by genetic transference	Usually remains dull and self-effacing	Tends to resemble male in appearance and voice
Effect on reproductive rate	Probably slight	In birds, limits size of brood	Permits raising of more young
Quality ultimately selected	Strength and vigor	Vigor	Cooperativeness, constancy

[1] From one point of view, this is also a mode of intrasexual selection, since the males compete with each other and may exclude each other from mating. But the distinctive feature of this mating system is free choice by the female, hence it is more appropriately called "intersexual selection."

inherit his size and strength, his belligerence, his antlers, horns, fangs, spurs, or whatever armament made him invincible. His daughters and granddaughters may, by genetic transference, bear his weapons, often in reduced form, or they may remain unarmed. Even if they are unencumbered by the essentially male characters, they should profit from the strength and vigor of their male progenitors, becoming better able to bear, rear, and protect their offspring. Confrontation of males for access to females preserves the vitality of a race, but not without the sacrifice of other valuable attributes.

Intersexual selection may be either unilateral or mutual. Unilateral sexual selection is found in its least diluted form in the courtship assemblies (frequently called leks or arenas) of male birds, where each calls and displays at his own station, with little or no interference by his neighbors, or at most ritualized confrontations. Since he offers to the female nothing but the insemination of her developing eggs, her choice is not influenced by anything except his personal qualities: his appearance, the intensity and persistence of his displays, his ability to win and hold a preferred situation in the assembly or to build an attractive bower. She is free to move from one to another of the males competing for her attention, perhaps joining each in a "dance" until she finds one who pleases her enough to win acceptance. This is the situation most favorable for the evolution of charming, gaudy, or extravagant plumage. Since the male will not attend nests or young, the development of his adornments is not limited by the need to become an efficient parent, not too conspicuous to approach the nest and not too encumbered with wattles or flowing plumes.

A psychic trait widespread in birds, as in humans, foments the tendency for adornments and displays to become extravagant. One might suppose that the expression of a character which for generations has contributed to the prosperity of a species would be preferred by each of its members; that any excess or deficiency in this character would be less acceptable. However, this is not true. As Niko Tinbergen demonstrated, some birds prefer exaggerated characters — "supernormal sign stimuli." Thus, an Oystercatcher (*Haematopus ostralegus*) will choose to incubate a clutch of five eggs in preference to its normal set of three; it will try to cover an artificial egg too big for it to sit on, neglecting its own egg nearby. A Ringed Plover (*Charadius hiaticula*), given the choice between its own lightly spotted egg and one with heavier markings, selects the latter. The Piratic Flycatcher (*Legatus leucophaius*), widespread in tropical America, builds no nest for itself but occupies the covered nest of some other small bird from which it has thrown out the builder's eggs or nestlings. If it finds a colony of oropendolas, it takes possession of one of the long, hanging pouches woven by these icterids. In these nests built by birds much larger than itself, the thief appears rarely to breed successfully, probably because the female becomes separated from her eggs in the litter of small leaves that covers the wide bottom of the pouch.

The more favored males in a courtship assembly may win a succession of females and sire many more young than would be possible for a monogamous male who helps to rear his nestlings. This reproductive advantage, in conjunction with the compelling effect of intensified characters, such as more brilliant plumage, more profuse ornamental plumes, or more spectacular displays, can lead to relatively rapid evolutionary change. Males may become burdened with ornaments that not only make them totally unfit for attending nests or chicks but appear also to impede their movements or annoy them. Examples include the Crested Argus Pheasant (*Rheinartia ocellata*), who bears in his tail the greatest of all feathers, six inches broad and over five feet long (15 by 150 centimeters); the peacocks' (*Pavo* spp.) heavy trains; the long, dangling wattles of the Three-wattled Bellbird (*Procnias tricarunculata*); the eighteen-inch (46-centimeter) ornament hanging from the breast of the Long-wattled Umbrellabird (*Cephalopterus penduliger*). A male bird who can forage adequately and escape enemies despite such encumbrances, like a man who can climb rocky peaks with a heavy pack on his back, must be an exceptionally fit and capable individual, who will transmit his vigor to his progeny of both sexes, although only his sons will inherit his ornaments, while his daughters remain dull and self-effacing. In effect, then, females who become entranced by the most lavishly attired males, or those who display most dashingly or persistently, choose for vigor, perhaps unwittingly.

Intersexual selection in its purest form, as exemplified by courtship assemblies, can be practiced only by species of birds of which the females can dispense with male assistance in rearing their young. This method of courtship occurs chiefly among nidifugous birds whose precocial young pick up their own food soon after they hatch, and among altricial birds who depend largely or wholly upon readily gathered fruits for nourishing the female parent and her brood.

The postures and movements of displaying birds are such as most effectively reveal their most brightly colored areas. We may ask whether the birds acquired the colors before the revealing attitudes, or the reverse. Almost certainly, the bodily postures or feather movements came first. Thus, excited birds frequently raise the feathers of their crowns. In many American flycatchers, the crown plumage spreads to reveal a small, usually hidden patch of brighter feathers, yellow, vermilion, or white. From such a modest beginning, royal flycatchers (*Onychorhynchus* spp.) developed their magnificent, wide-spreading scarlet aureoles, bordered with violet spots, which they unfold far too seldom to satisfy their admirers. In a threatening attitude, pigeons often lift their wings, which can deliver strong blows. The undersides of the wings of White-tipped Doves (*Leptotila verreauxi*) and related species have become cinnamon, the brightest color on these plainly attired pigeons.

A fact often overlooked in discussions of sexual selection is that the situation among migratory territorial birds in spring has much in common with that of permanently resident birds, such as certain birds of paradise and cotingas, whose displaying males, instead of being closely aggregated in a typical courtship assembly, are more widely scattered in a dispersed, or "exploded," lek. Although the mating of the former is usually treated under the rubric of pair formation, they provide good examples of unilateral intersexual selection. The males of these migrants nearly always arrive from their winter homes at lower latitudes before the females. After claiming a territory, each male sings to advise neighbors that he is in possession and to advertise his availability to females, who are free to choose the partner who will share their parental chores.

Although a female at a typical courtship assembly need consider only the personal attributes of the male who will father her nestlings, a female choosing a territorial male does well to pay attention also to the quality of the territory he offers, whether it includes adequate sites for her nest and is likely to yield enough food for parents and young. On an islet in Japan, Hisashi Nagata found that females of the monogamous Middendorff's Grasshopper Warbler (*Locustella ochotensis*) chose males with the largest, most productive territories; between the plainly colored males themselves, he could discover no consistent differences that might influence the females' preferences. Females of predominantly monogamous species, including Dickcissels (*Spiza americana*), Bobolinks (*Dolichonyx oryzivorus*), and Eastern Meadowlarks (*Sturnella magna*), sometimes choose an already-paired male with a superior territory, although bachelor males with less productive territories are available to them, thereby giving rise to simultaneous polygyny. The two, rarely more, females of the polygynous male may divide his territory between them. He may help both to feed their young, giving preference to his first, or primary, mate.

A chief difference between the situation at courtship assemblies and that among territorial birds is that among the latter the female's choice of a partner is complicated by the quality of his territory or, in the case of hole-nesting birds, the cavity he offers to her. Another difference is that a dominance hierarchy is less likely to develop among birds spread over rather extensive territories than among birds concentrated in an assembly. Accordingly, males are less likely to display their adornments to each other. Their interactions more often take the form of counter-singing, with a certain amount of skirmishing at the territorial boundaries. We are left with female choice as the most probable explanation of the beauty of the males in the many species in this category that exhibit pronounced sexual differences. The degree to which female preference has improved the quality of the males' singing is more problematic. Song serves to advise neighboring males that a territory is occupied and, moreover, must be distinctive to guide arriving females to their own rather than another species.

Despite the resemblances between male selection in migratory, monogamous, territorial birds and that of nonpairing males in courtship assemblies or more widely dispersed, males of the first category do not develop the extravagant adornments of many males in the latter category, for two good reasons. The first is that the monogamous males, however colorful they may become, must avoid structures that would make them less competent to feed the large broods that many of these migratory birds must rear to compensate for the numerous hazards to which they are exposed but from which permanent residents of low latitudes are exempt. The second is that plumage which broke the streamlined contour of their bodies would be a great impediment on long migratory flights and, if molted before the journey began, would be expensive to replace in a spring crowded with other energy-consuming activities. Moreover, since a monogamous bird cannot engender as many offspring as one who forms no lasting attachment, the reproductive potential that favors the evolution of the most embellished males is lacking. Exceptional among monogamous, nest-attending males is the Resplendent Quetzal (*Pharomachrus mocinno*), a tropical bird who performs only short altitudinal migrations. Despite his ornamental plumes, he takes a large share in incubating the eggs and feeding the young, to the detriment of his very long upper tail coverts, which are badly frayed, if not broken off, by friction and bending, while he broods and nourishes his nestlings in a deep cavity in a tree.

Animals that establish enduring monogamous bonds often unite in a manner less spectacular than those that we have already reviewed. Frequently, neither sex obviously takes the initiative; a male and female appear to forge a bond by mutual consent after tentative approaches and trials of compatibility. Since monogamy prevails among birds but is practiced by only a minority of mammals, it is chiefly among the former, especially those permanently resident in mild climates, that we find mutual intersexual selection, which for brevity we may call mutual selection. Often when only a few months old, birds form pairs that will not nest for many additional months, sometimes more than a year. Of all mating systems, intersexual selection, especially as practiced in leks, consistently promotes the most striking contrasts between the sexes in appearance and behavior, whereas mutual selection leads not only to the most enduring pair bonds but likewise the greatest similarity. Often the male and female of a pair so closely resemble each other that they can be distinguished only by behavior or dissection. After mating, large birds like grebes and penguins continue to engage in elaborate displays in which the two partners play identical roles; or they reverse roles, even to the extent that the female sometimes mounts the male. Such spectacular epigamic performances help to bind the two together, to synchronize their readiness for coition, and perhaps to mitigate the effects of harsh weather at high latitudes.

During the long season when their reproductive organs are shrunken and dormant, many birds permanently resident in the tropics live two by two. Sometimes these pairs are evident even in large flocks, like those of macaws and smaller parrots, in which mated birds fly close together, with wider separation between themselves and other flock members. When partners come together after a brief separation, they may greet one another with fluttering wings and/or special vocalizations seldom heard on other occasions. Where constant visual contact is difficult, as among wrens that forage amid dense vegetation, the mates keep in contact by song. Dueting is frequent among constantly mated birds. Not only in plumage but likewise in voice mutual selection tends to make the sexes similar, although the female's song may be shorter and her voice somewhat weaker than her mate's. As the nesting season approaches, the male often feeds his partner, thereby helping her to form her eggs. He may join her in choosing the nest site, or sometimes take the initiative. He may or may not help to build, and less often, especially among passerines, he incubates, but nearly always he helps to feed and protect the young.

It has been claimed that sexual selection is simply one of the many ways in which natural selection operates. Sexual selection is a mode of natural selection only inasmuch as everything that occurs in nature is natural. It differs profoundly from the widespread modes of natural selection, such as predation, disease, starvation, and environmental extremes. These eliminate the less fit while offering no special advantages to the more fit, except perhaps, indirectly, reduced competition for resources. Natural selection is essentially negative selection and were better called "differential survival and reproduction." In sharp contrast to this destructive selection, sexual selection is positive, choosing favored individuals to contribute to the ongoing life of their species, to become collaborators and constant companions, without (except in the cruder forms of intrasexual selection) injury to individuals who fail to win partners. If they escape the rigors of natural selection, many of the individuals who fail to mate in one season may achieve parenthood in a later season, when they are older, with more elegant plumage or more skillful or persistent displays. Far more than natural selection, sexual selection resembles the methods by which horticulturists "improve" ornamental plants, and often with a similar result, increase of beauty.

Moreover, sexual selection and ordinary natural selection often pull in opposite directions. The former frequently promotes brilliant plumage, extravagant adornments, noisy and conspicuous behavior, all of which make birds more vulnerable to predation. They consume a bird's energy while they may impede his foraging, especially if he bears very long plumes or clumsy wattles. Natural selection tends to make animals subject to predation (as most birds are) inconspicuous, conservative of energy, and streamlined for more effective action. The appearance and behavior of many male birds appears to represent a state of equilibrium between these opposing selective agents. In tropical lands where seasonal differences are relatively slight, this balance is usually maintained throughout the year; few birds have strikingly different breeding and off-season plumages. In regions with extreme seasonal contrasts in climate, as in much of the North Temperate Zone and the Arctic, the balance is, for many species, preserved only during the breeding season. After its termination, many of the more brilliant males molt into a much duller plumage, in which they either migrate to milder climates or remain to endure a harsh winter. The contrast between the breeding plumage, a product of sexual selection tempered by natural selection, and the winter plumage, subject to natural selection alone, reveals vividly the opposite effects of these two modes of selection.

Predators, those major agents of natural selection, care little for the personal qualities of their prey, other than size, availability, and perhaps taste. Except when they live in pairs or family groups, like parrots or geese, birds that flock seem rarely to have particular companions; like Sanderlings (*Calidris alba*) on the seashore, they are attracted to their species but not to individuals. Sexual selection provides our earliest clear examples in the animal kingdom of the selection by one individual of another for personal qualities such as appearance, behavior, and prob-ably other attributes that we fail to recognize. It is an important step in the emergence of personality from the level of specific uniformity. When mutual, sexual selection leads to lasting individual attachments and, ul-timately, to friendship and conjugal fidelity, thus contributing to moral as well as physical beauty.

References

Darwin 1871; Huxley 1938a; 1938b; Myers 1983; Nagata 1986; Tinbergen 1951.

6.
The Courtship of Grouse, the Great Argus Pheasant, and the Ruff

As our first example of a bird that courts in assemblies, let us take the Black Grouse (*Lyrurus tetrix*), widespread in northern Eurasia and in alpine regions farther south. The male, about twenty-one inches (53 centimeters) long, is black glossed with dark blue, with a white wing-bar and undertail coverts. The feathers of his long, forked tail bend outward on each side in a graceful lyrate curl. Above each eye, he wears an inflatable red comb. The smaller female and juveniles of both sexes are brown, with blackish mottling above and blackish bars below. Their forked tails lack the lyrate curve. On moorlands, heaths, meadows, and other open spaces with scattered trees or bushes, they eat buds of broadleaved and coniferous trees, foliage, and berries. Chicks prefer insects.

Throughout the year, adult males live in small groups; females without young are more often solitary. The sexes are seen together only on their display grounds in the mating season. For their nuptial displays, black cocks choose open moors, bogs, or meadows with wide visibility, but often with nearby trees or shrubs, on farms, near buildings and roads, as well as in wilder places. If not seriously disturbed, the same display area is occupied year after year by up to twelve or more cocks, who appear to form a social group. Occasionally a male displays alone. On these assembly grounds, individual cocks defend territories that vary greatly in size according to the terrain and the number of birds in the gathering, from about thirty-five to one thousand square yards (30 to 836 square meters).

In the frigid darkness of northern lands before daybreak in early spring, the black cocks announce their presence in the trees around their leks by repeating a musical *roo-koo*, audible from a quarter to half a mile away. While it is still too dark to see more than their white undertail

Black Grouse, *Lyrurus tetrix,* male. Northern Eurasia.

coverts, the dusky birds become active on the ground. Through the early morning, their displays continue, to be resumed, after a long day spent eating and resting, on a minor scale in the late afternoon, just before the birds go to roost. A cock repeats his *roo-koo,* loudest of his utterances, mainly from his own territory, thereby advertising his presence to the hens. While calling, he bows low, with his swollen neck stretched forward. When a gray hen flies over the assembly, the males often "flutter jump" a few feet into the air, with clattering wing-beats as they rise and fall. After alighting, facing the direction taken by the hen, they stretch up their heads and crow with a harsh, blowing or hissing *tshoo-wooeesh.* The sound of a hen taking flight with loud wingbeats often stimulates the cocks to flutter jump before she becomes visible. Any sudden noise, such as explosions at a distant quarry, the barking of a dog, or the cackling of a domestic hen, may set off the flutter jump and crow.

Each Black Grouse zealously defends his space from adjoining males. When two confront each other at their common boundary, their interactions are more formal than fierce. Although mostly they *roo-koo* within their own territories, turning to face in various directions, sometimes one advances, calling, toward a neighbor. The latter runs to meet him with short, rapid steps, puffed-out neck stretched forward, wings and tail fully displayed, silently, or voicing a rattling *ca carrr.* The interaction that follows reminds one of the confrontation of two male songbirds at their territorial boundary. While they stand face to face, one black cock retracts his head as the other darts his head forward. Then, as the protruding head is withdrawn, it is followed closely by the advancing head of the other, much as in some mechanical toy. Or the two may alternately advance and retreat over a short distance. With intervals during which the opponents may do no more than face each other in full display, these formalities may continue for many minutes, at the end of which, as by common agreement, the actors separate and return to their own places. Occasionally, however, they fight, with fluttering wings, leaping up at each other and striking out with their spurless feet. Interlocked, they may tumble together to the ground; rarely one is killed. Formal encounters greatly reduce the incidence of serious fighting, which could not become frequent without disrupting the communal courtship system.

When a gray hen walks into a lek, each cock, as a rule, displays to her on his own plot of ground and refrains from following her if she passes into another's domain. The male favored by her presence greets her with a *roo-koo* while he faces her, then courts her, much in the manner of a pheasant, with a lateral display. Tilting his whole body sideward and downward, drooping and spreading his nearer wing, he presents to her as much of himself as he can make visible in a single view as he circles around her, first on one side, then the other, without passing behind her, where he would not be visible. She sees his head with engorged, swollen crimson combs; his glossy, distended neck, depressed before her; his uniquely curled tail. When he passes in front of her, his white undertail coverts contrast with his black tail feathers. If she advances a few steps, he runs ahead of her. He repeats his tilted circling before her, showing all his special adornments. She appears attentive to his elegance, almost self-conscious and nervous. This lateral, tilted nuptial display differs from the frontal aggressive display of rival cocks, which does not so completely reveal the male's adornments.

Despite splendors so carefully paraded before her, the cock may fail to win the hen, who passes onward to a neighbor; he cannot mate with her unless she consents. Sometimes, failing sufficiently to impress the hen, the cock circles until he faces her and prostrates himself in front, with his head held up and his fanned-out tail erect, as though suggesting to her the posture she should assume for coition, or beseeching her to comply. The hens appear to have definite preferences. One sometimes attacks another who interferes while she is being courted, or chases her away more swiftly than the cocks often run. A hen sometimes interposes herself between two fighting males.

When a hen crouches in front of a displaying male, he mounts her, fluttering his wings to keep his balance and with his bill holding the feathers on the back of her head, much as a domestic rooster does. Observers differ about the behavior of neighboring males while coition occurs. E. O. Höhn noticed that each remained at his own post during the act, as is usual in avian courtship assemblies. Edmund Selous saw other cocks rush to interfere with the mating pair, who were spared from rude interruption by the brevity of coitus — lasting about ten seconds — in relation to the distance separating the cocks on their territories. In any case, after her partner dismounts, the hen preens her plumage and marches away, to lay her eggs and rear her chicks with no help from him. On his part, the successful Black Grouse may promptly resume rookooing, or court another hen who has entered his territory. One was seen to circle around three females at the same time. Postnuptial displays are absent.

The Black Grouse's vernal displays cease after all or most of the gray hens have been inseminated and laid their seven to ten eggs in a sparsely lined nest well hidden on the ground. Months later, about mid-September, the cocks resume their displays on the same areas that they occupied in spring, and continue until October's end, occasionally longer. In autumn the evening exercises of springtime are omitted and the birds perform only early in the morning, less vigorously than they did during the lengthening days. Roo-kooing, crowing, and threat displays now occupy them. These autumnal activities attract females who perch beside the leks or fly low above them, without walking through them, as they do in spring. They are not courted at this time.

The courtship assemblies of the Black Grouse give play to both intrasexual selection (the cocks vie with each other and occasionally fight) and intersexual selection (the hens freely choose their partners). From the detailed accounts of Selous and Höhn, I judge that intersexual selection predominates, and female choice has promoted male adornment.

Over the vast sagebrush plains of the northwestern United States and extreme south-central Canada, Sage Grouse (*Centrocercus urophasianus*) roam through much of the year in large flocks of both sexes, eating sagebrush leaves, their principal food. The biggest member of the grouse family in North America, the male is about thirty inches (76 centimeters) long. Clad largely in brown or gray-brown flecked with white, he has a black foreneck and belly and a white breast. His eighteen tail feathers taper from broad bases to long, attenuate ends; his dark undertail coverts are tipped and spotted with white. The much smaller female is more uniformly grayish brown, with fine buffy and whitish mottling.

Sage Grouse, *Centrocercus urophasianus,* male displaying.
Northwestern United States and south-central Canada.

In late February or March, the birds gather on their strutting grounds, or display areas, on open plains or gentle slopes covered with short grass surrounded by sparse, low sagebrush. The display area may be as small as one acre (0.4 hectare) or as large as forty acres (16 hectares) and, at the peak of the season, attended by as many as four hundred males and about as many females. Year after year, the birds assemble in the same place for their nuptial exercises. After spending the day eating and resting in neighboring or distant areas, the cocks gather in the late afternoon on the display ground, to occupy their accustomed spots, challenge one another, and occasionally fight. The arrival of a few hens at this time intensifies these preliminary exercises. When night approaches, the birds leave, singly or in groups, to roost nearby or several miles away. As the new moon rises higher in the sky at nightfall, more and more cocks remain near or on the display ground, where they have been seen strutting, challenging, and fighting in the middle of the night. Hens are sometimes present on the moonlit arena.

The grand period of activity begins with the first promise of daybreak when, early in the season, frost or snow often covers the ground. The principal activity of the male Sage Grouse is strutting. He draws himself upright, erects and spreads his tail with the attenuated feathers spread out and widely separated, like the rays of a halo. His wings are raised at the base and bent sharply downward at the wrist, the tips of the longest primaries often touching the ground. By swallowing air he fills his air sac, an expansion of his esophagus, until it swells out hugely with four or five quarts (liters) of air, spreading the stiff white feathers of his breast until they cover the whole front of his body and hide his head. In the midst of this white expanse appear two egg-shaped patches of yellowish bare skin. While he inflates this pouch in stages, he advances three or four steps and turns ninety degrees or more. No sooner has he puffed out his foreparts to their fullest than he forcibly expels the air, while his breast contracts and the bare spots shrink until they vanish. The expulsion of the air is accompanied by a *plop* audible half a mile away on a still morning. After deflation, the cock pauses to look around briefly and observe the effect of his display before he resumes it. As long as hens are present on the lek, the cocks continue to strut, with their stiff attitude and grotesquely swollen bodies appearing more bizarre than beautiful.

For several weeks before the hens arrive, the cocks assemble on their display ground and by strutting, threatening, bluffing, and fighting establish individual territories and determine the rank order that will affect their chances of winning hens. A challenging Sage Grouse runs toward another with guttural, menacing notes. Often only a few wing blows are exchanged before one admits defeat and retires into his own territory, where he can nearly always hold his own against a bigger bird. If the opponents are more evenly matched, they stand side by side, head to tail, a foot or more apart. With body, wings, and tail quivering with excitement, they rapidly repeat the guttural challenge. Suddenly, one lashes out with a wing at the other, who may dodge or parry the blow and strike back in turn. Rarely, one seizes with his bill the top of the other's head and holds him while thrashing him loudly with a wing, despite his struggles to escape. More often, before the fight escalates to this extreme, one of the contestants slowly backs away after the exchange of a few blows. After the cocks separate, each may continue to strut in his own small territory with an area of 16 to 120 square yards (13 to 100 square meters), which he occupies every day.

By these confrontations, the cocks learn their relative strengths and the organization of the courtship assembly is established. In a central position called the mating spot (of which as many as six may be present on an extensive lek) stands the alpha or master cock, usually surrounded by a compact group of hens, all in a space about the size of an ordinary room. Nearby struts the beta cock, the master cock's chief rival. Around these two are grouped subordinate cocks, usually three to six in number, called guard cocks because they prevent males still lower in the hierarchy, mostly younger individuals with peripheral territories, from intruding upon the central two. The guards strut frequently and are tolerated by the dominant birds if they do not come too near and mingle with the hens. The master is not always the biggest bird in his group, but he is one of its older members, active and pugnacious, ready to defend his coveted station against all pretenders. The beta cock and guards are large and apparently also older individuals.

Hens first appear on the strutting ground two or three weeks after the cocks take up their stations and begin to perform there in March. After they begin to attend, they arrive in the morning from ten to thirty minutes later than the cocks. Many walk into the lek; those who come flying seldom alight at its center. As they stroll through the assembled males, they appear unmoved by all the displays that they incite. They may wander about the lek, stopping here and there to pick up food, to rest, or to look around. They may pause near a cock who impresses them, without inviting him to mount. If not yet ready to mate, the hens march away after their visit of inspection, perhaps to return on the next morning. The cocks do not molest them as they depart. If prepared to mate, they gravitate to the strutting master cock, who at the peak of the season may be closely surrounded by from fifty to seventy of them, waiting for his attention. Although on the whole they quietly await their turns, occasionally preening or squatting down to rest, sometimes, growing impatient, a hen pecks another or drives her from the master cock. Once a thwarted hen pecked vehemently at the cock, who showed no resentment. A neglected hen may call plaintively *quer quer quer,* as though complaining or begging for the cock's attention. If kept waiting too long, one hen sometimes mounts another, or three or four may pile up in a heap, all attempting coition in the wrong context. And all this while, many surrounding males are eagerly waiting to serve these same frustrated hens!

When her turn comes, the hen squats in front of the cock with out-spread wings usually touching the ground. He mounts her and in a few seconds completes coition. She rises quickly, runs a short distance, ruffles her plumage, and vigorously shakes it. After a bout of preening, she walks, or occasionally flies, from the strutting ground, to lay one of her seven to nine eggs, rarely more, in a slight depression lined with dry grass and leaves, on the ground in the shelter of a sage bush or tussock of grass. Without a male's help, she incubates her eggs and rears her chicks.

Most matings occur before sunrise, when Golden Eagles (*Aquila chry-saetos*), the Sage Grouse's most dreaded enemies, seldom fly over their strutting grounds. On a single morning, a master cock may mate with up to thirty hens. One morning, a head cock covered nine hens in thirteen minutes. But the male Sage Grouse's sexual capacity is not unlimited; eventually he becomes satiated or exhausted, makes abortive attempts to mate, or neglects a hen soliciting his attention, perhaps fighting another cock instead of mounting her. The master cock's exhaustion is the beta cock's opportunity, but he may be attacked by the former while he tries to mate. At times the subordinate charges his superior after the latter has mated with many hens. Of a total of 174 matings observed by John W. Scott in 1941, 20 were by cocks of undetermined rank. Of the remaining 154 matings, 114 (74 percent) were by alpha cocks, 20 (13 percent) by beta cocks, 5 (3.24 percent) by guard cocks, and 15 (9.78 percent) by isolated cocks outside the regular mating spots or at hurriedly impro-vised ones. Similarly, R. Haven Wiley found that in each of three years fewer than 10 percent of the males completed more than 75 percent of all matings. Cocks low in the hierarchy have a chance to mate as hens who have not been served by the busy principal cocks walk away when the assembly is about to disperse late in the morning, or after the peak of the season, when the master cocks and sub-cocks have been surfeited. On the Laramie Plains of Wyoming the courtship assembly, started in early March, dissolves around mid-June. Evidently, if all goes well, each female raises a single brood each season.

The observations of Scott in Wyoming and Wiley in Montana sug-gested that in the Sage Grouse parenthood is determined largely by the contests of males for dominance and possession of the privileged mating spots in the strutting grounds, therefore by intrasexual selection. Fe-males appeared to accept the outcomes of these contests, choosing the most vigorous and aggressive cocks to sire their chicks. Later studies by R. M. Gibson and J. W. Bradbury at smaller assemblies in eastern California modify this picture. Here, in one instance following a severe winter and occasionally at other times, territories were abandoned and matings were not concentrated at central spots. In some years, males frequently left their territories to mate with females outside the lek. The hens' choice of cocks was influenced by the quality and frequency of their strut displays more than by their location in an assembly. Intersexual selection prevailed; the hens chose freely, and their preference for males with exaggerated secondary sexual characters has helped to make the puffed-up, pugnacious male Sage Grouse more grotesquely imposing than graceful and lovely.

Other members of the grouse family that court in assemblies are the three species of *Tympanuchus,* the Greater Prairie-Chicken (*T. cupido*), Lesser Prairie-Chicken (*T. pallidicinctus*), and Sharp-tailed Grouse (*T. phasianellus*). The first of these is a bird about eighteen inches (46 centimeters) long, barred with dark brown and buffy white in a bold, zebralike pattern, with a short brown tail. When displaying, the male engorges the yellow combs above his eyes and inflates an air pouch on each side of his neck until it looks like a bright orange fruit, shading to red at the top. At the same time, he erects long eartufts that rise above his hindhead and appear to balance his upturned tail. Formerly widespread over the eastern and central United States and adjoining parts of Canada, Prairie-Chickens now occupy a shrunken range on the grasslands of north-central United States and central southern Canada. Here at dawn on spring mornings, and less frequently in the evenings, the males gather to attract the hens. Each has his own station separated by about ten yards (9 meters) from those of his neighbors. Cooing or booming all together, the birds in a large gathering keep up a continuous roar audible afar. Before booming, the cock performs a little dance, his pattering feet beating on the ground a tattoo which does not carry nearly as far as the loud booming.

Observers have apparently given more attention to the sounds and fights at the Prairie-Chickens' leks than to their mating. Suddenly, with henlike cackles, a cock springs a foot or two straight upward and whirls about. Then, with head down, combs and neck pouches inflated, ear tufts and tail erected, one advances by little runs toward a neighbor, who meets him with the same display. Uttering low, whining notes, they fight so violently with wings, bills, and feet that feathers strew the ground. Sometimes one chases another for many yards.

Other members of the grouse family display singly rather than in as-

Greater Prairie-Chicken, *Tympanuchus cupido,* male displaying.
North-central United States and south-central Canada.

semblies. In addition to diverse vocalizations, to spreading tails, wings, and capes, and to the inflation of colorful combs and neck-sacs, their performances include the wing-drumming of the Ruffed Grouse (*Bonasa umbellus*) on a log or rock in a northern woodland, the wing-clapping flight of the Spruce Grouse (*Dendragapus canadensis*) in a coniferous forest, and the song-flight of the Rock Ptarmigan (*Lagopus mutus*) above the Arctic tundra.

Omitting, to avoid prolixity, the Lesser Prairie-Chicken, the Sharp-tailed and other grouse, and the ptarmigans of northern lands, we turn to a tropical bird, the Great Argus Pheasant (*Argusianus argus*), a brown bird with a huge tail that makes the male about six feet (183 centimeters) long. In evergreen forests of the Malay Peninsula, Sumatra, and Borneo, these terrestrial pheasants roam singly over large home ranges, picking up fallen fruits and scratching in the ground litter for insects. Solitary adult and subadult males reveal their presence by long calls of fifteen to seventy-two hoots, so loud that they can be heard more than half a mile (1 kilometer) away, and are answered by other males. Unlike many other

Ruffed Grouse, *Bonasa umbellus,* male drumming.
Canada and northern United States.

pheasants, Great Argus males lack spurs on their legs and appear to interact with each other by voice alone. They acquire adult plumage when about two years old, but their wing and tail feathers grow longer with each successive molt until the sixth or seventh year. They may live for more than twenty years.

When they start to clear their courts, in January and February in Malaya, adult males change from long to short calls, each a burst of one to twelve high-pitched hoots that sound like *kaw-wow.* Courts are usually situated on or near the crest of a wooded ridge or other elevation, which increases the distance that the calls can be heard. Each court is a patch of ground from which the pheasant tosses aside with his bill, or fans

Great Argus Pheasant, *Argusianus argus,* male.
Malay Peninsula, Sumatra, Borneo.

away with flapping wings, all leaves and other litter. He pecks and pulls to uproot small obstructing plants, until his court appears to have been swept clean with a broom. The shape of this bare area may be roughly circular or irregular, its outline determined by obstructions such as trees or termite mounds. Small at first, it gradually expands during the season of courtship until it may become from fourteen to twenty square yards (12 to 18 square meters) in area. One exceptionally large court covered eighty-six square yards (72 square meters). While alone at his court, the male rests on a low branch or slight rise of the ground beside it and repeats short calls every few minutes. At intervals, he preens, or jumps down to clear his stage. Or he rushes all over it in an erratic course, jumps, flaps his huge wings, and erects them into a fan more than a yard wide, all with pauses to call. If he did not prepare a clear court, a bird two yards long could hardly perform in this wild manner without becoming entangled in the forest undergrowth.

In 191 days of observation, G. W. H. Davison saw a female visit a male's court only once. Birds as solitary as Great Argus Pheasants do not approach one another without initial mistrust. A female may enter a court so aggressively that its owner stands aside. As she calms down, he marches around her in a wide circle, head bent low, rhythmically stamping the bare ground. Spiraling inward until close to his visitor, he halts, and changes from the lateral display widespread among pheasants to frontal display. He spreads the broad secondary feathers of his wings until they surround his body like a great fan, from the top of which project two long tail coverts with scalloped inner edges. Staring at the female from the surface of this fan are hundreds of eyespots, whose realistic ball-and-socket appearance stirred the wonder of Darwin. With his head hidden behind his expanded plumage, the displaying argus peers out at the visitor through a narrow space at the base of a wing. Eyes or their representations arrest the attention of birds and other animals; the female pheasant could hardly fail to be impressed by the marvelous array of them that the courting male turns toward her. After mating, she departs to lay two eggs on dead leaves on the ground. For twenty-four or twenty-five days she incubates continuously, never leaving the nest to eat or drink but sitting with a constancy equaled by few other birds. She alone attends her chicks, at first feeding them from her bill, as few gallinaceous birds do.

Whereas the grouse and sandpipers that appear in this chapter gather in assemblies, the courts of the Great Argus Pheasant are widely separated, the average distance between them being about thirteen hundred feet (400 meters). A male at his court can often hear the loud hoots of several others; but females, who travel chiefly on foot, could hardly visit a number of males in a short interval. During his three-year study of argus pheasants, Davison found no evidence that females wandered from court to court. It would be difficult for her to compare the adornments of different males, as females of species with compact courtship assemblies can readily do. But, unless females choose males with the most abundant or realistic ocelli, how can we account for the evolution of these wonderful designs? Could this have occurred in past ages when Great Argus Pheasants were more numerous and their courts closer together? Humans have long preyed upon these large birds. Natives of Borneo set bamboo stakes in the pheasants' courts. While struggling to remove these intruding objects, the birds often cut their throats on the sharp upper ends and perish by bleeding. Now they are shot with guns.

Another pheasant that clears a court on the ground in forests of southeastern Asia is the Crested Argus (*Rheinartia ocellata*). Feathers of both argus pheasants have been found on the same court, suggesting that occasionally one may use a bare patch cleared by the other. When a female visits the court of a Crested Argus Pheasant, he displays laterally to her, spreading like a sail his enormous tail feathers, the largest worn by any bird. I have found no study of the courtship of the Crested Argus in its native forests. So, for the present, we must turn aside from this magnificent family, which includes peacocks, peacock pheasants, Golden Pheasants (*Chrysolophus pictus*), Amherst Pheasants (*C. amherstiae*), tragopans, and many others, whose ample plumage is adorned with more intricate details than the most ornate small bird can display.

Most sandpipers are plainly attired and difficult to identify in plum-

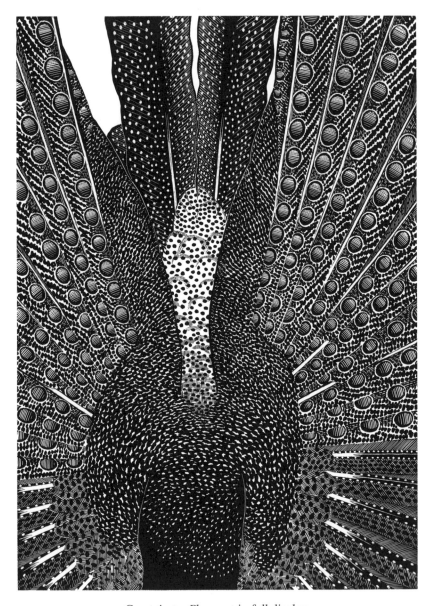

Great Argus Pheasant in full display.

Lady Amherst's Pheasant, *Chrysolophus amherstiae,* male.
Southeastern Tibet to upper Burma.

age streaked, spotted, barred, or marbled with grays, buffs, black, and white. Their utilitarian dress lacks ornamental plumes. An outstanding exception is the male Ruff (*Philomachus pugnax*) in nuptial array. About eleven inches (23 centimeters) long, he is adorned with elongated facial tufts that when erected rise above his head like a mane, and an ample erectile ruff that covers his chest and shoulders and extends behind his eartufts. Usually the ruff and earcoverts are differently colored; either may be black, blue-black, some shade of brown, scarlet, orange, or white, and plain, barred, spotted, or blotched. His remaining plumage, also quite variable, is finely vermiculated, or boldly barred with black or gray and buff. The bare, finely papillate skin of his face may be gray, yellow, orange, or scarlet. The bill and legs show an equal diversity of colors. Males in cryptic winter plumage and females at all seasons are as plain as most sandpipers. Not only do male Ruffs differ from all other members of the sandpiper family in their profuse nuptial attire, they display, in a single population, individual diversity unequaled by any other undomesticated bird. How can we account for this amazing variability?

Perhaps a study of the Ruff's social system will throw light on this

Ruff, *Philomachus pugnax,* males. Northern Eurasia; in winter, from
southern Eurasia and Africa to Australia.

problem. Returning in spring from their wintering grounds in southern
Eurasia, Africa, Indonesia, and Australia, Ruffs establish their leks on
low meadows over much of northern Eurasia. The site of such an assemblage is occupied by a succession of birds over a period of many years;
even the opening of a road through the midst of one of these display
grounds did not cause abandonment. Each display area is dotted with a
number of circles, each about one foot (30 centimeters) in diameter,
from which the flowery meadow herbage has been worn away by the
constant trampling of the Ruffs, each on the little bare plot that he claims
as his own. These courts are usually from twenty to sixty inches (50 to
150 centimeters) apart.

Long before sunrise on bitterly cold April mornings, "whilst learned
ornithologists, all the world over, lie sleeping in their pleasant beds,"
Edmund Selous, a pioneer student of avian courtship, hid himself in
view of a lek that was sometimes occupied by a score or more of Ruffs,
no two of which were alike. As day grew brighter, a sort of madness
periodically seized the whole assemblage of Ruffs, who rushed about,
darting and springing and kicking and whirring their wings like frantic
creatures. Such bursts of maximum excitement were brief, rarely lasting

more than three to five seconds. At times, especially when a reeve (as the female Ruff is called) arrived, two Ruffs would rush and leap high in the air against each other, in a momentary clash that was scarcely ever renewed after they descended. Sustained duels were rare and never ended in injury, as far as Selous could learn. A cause of disturbance was the arrival of a Ruff who alighted not quite within his own precincts, displacing a neighbor who started a skirmish that spread through the crowd. However, such flare-ups were inconsequential, for, as Selous insisted, the assembly of Ruffs was dedicated to the service of Venus, not of Mars.

The approach of a reeve stirred the strongest excitement, which began before she touched the ground. The waiting Ruffs darted wildly about, sometimes colliding and facing each other belligerently for a few seconds. Except for the whir of wings and the frequent dull thud of feathered bodies colliding, the birds were silent; rarely did they use their voices in these gatherings. Suddenly, all dropped to the ground, each in his own little territory, where he lay crouching, legs folded beneath his body, head extended forward, feathers ruffled, wings half open, tail spread and depressed. Thereupon, the reeve, who always seemed calm, unperturbed by this maelstrom of passion and desire into which she had intruded, walked deliberately up to the prostrate Ruff of her choice and pressed close to him or fondled his feathers with her bill. Then she in turn crouched, perhaps to receive similar caresses from him before their union was consummated. Rarely did neighboring Ruffs attempt to interfere in the nuptial rites. On the few occasions when a too-eager Ruff tried to force the compliance of a reeve, he failed. The Ruffs seemed aware that without her approval and cooperation they could accomplish nothing, and as a rule they waited until she had signified her choice.

It was plain that in this assemblage, which at times contained as many as twenty-two males, the females preferred certain individuals above others. One of these favored males had an exceptionally large, soft, thick, light golden brown ruff and darker brown head lappets. The other male had a blue-black ruff and flowing white head tufts tinged with brown. These were the two birds whose appearance Selous most admired; but aside from the fullness of their plumage, they did not appear more vigorous than less successful competitors. The more frequently chosen of these two was the brown bird, who showed the greatest development of the feature that is most characteristic of the species, and responsible for its name. The fact that the Ruffs whom Selous most admired were also preferred by the reeves suggests a fundamental similarity in the visual perceptions and aesthetic responses of birds and humans.

Since Selous made his pioneer observations on the courtship of Ruffs early in this century, others have studied them in greater detail by more modern methods. Some of the results of these more recent studies have been summarized by Julia M. Shepard, who herself contributed substantially to our understanding of their behavior. She showed that males do not consistently retain the same bare courts throughout the mating season; shifts in occupancy occur, one male sometimes displacing another. A peculiar feature of these courtship assemblies is the presence of satellite males, who are distinguished from the others by their usually white or light-colored ruffs and head tufts. Lacking a court of his own, a satellite attaches himself to a territorial male and temporarily shares his court, where he is tolerated because he is not aggressive. After an interval there, he may join a different male in the same lek or move to another lek.

In a painstaking study continued for three seasons, Shepard tried to unravel the factors that determine a reeve's choice of a Ruff. She discovered no single dominating factor but rather an unstable constellation of factors that varies from lek to lek, and from day to day at the same lek, as the behavior and physical state of the component males change with the advancing season. A reeve entering an assembly for the first time walks around, probably assessing the different stimuli provided by the males on their several territories. She may visit the lek repeatedly before she meets a situation that incites her to solicit a male. Like a Sage Grouse, she may prefer a male near the center of the assembly. Males who display more actively are likely to attract her, but excessive activity may frighten her. The presence of a satellite on a male's court may weigh in his favor. The male's appearance, his ruffs and tufts, which increase in length and possibly in profusion with his age, is not without influence on a female seeking a partner. These factors, singly or combined, may stimulate different females to varying degrees. Evidently, inviting a male to mate with her is not such a simple matter as finding a single effective "releaser." In any case, after her eggs are fertilized, the reeve goes off, lays the sandpiper's usual four in a simple nest on the ground, and has no further need for a male until the following year.

The complexity of the factors influencing the reeve's selection of a nuptial partner appears to throw light upon the extraordinary variability of the Ruff's plumage. If the appearance of the males were the only, or the weightiest, factor determining her choice, we would expect them to be as uniformly colored as most male birds in full breeding plumage. But the subtle interplay of determining factors leaves this character free to vary in a more or less random manner. If reeves did not prefer males with ruffs and tufts, Ruffs would probably lack them. But the females' preference for any particular color is not strong enough to outweigh certain other advantages that individual males may have, and stabilize the coloration of these adornments.

Like the Sage Grouse, male Ruffs sometimes mate with females away from courts.

References

Bent 1932; Davison 1981; Gibson and Bradbury 1987; Höhn 1953; Lank and Smith 1987; Scott 1942; Selous 1927; Shepard 1976; Wiley 1973.

7.
The Courtship of Hummingbirds

Hummingbirds are the sparkling gems of the feathered world. No other birds concentrate so much brilliance on such small bodies. Although a minority are mainly brownish or bronzy, they are predominantly shining green, which is nearly always relieved by other bright colors. On many species the throat or crown, or both, gleam with an intense metallic refulgence, magenta, ruby, sapphire, violet, purple, or green. Often this covers an expansible gorget. Hummingbirds owe their glittering iridescent colors not to pigments but to the internal structure of their feathers, which produces them by such optical phenomena as reflection, refraction, and interference. These structural colors, like those of the rainbow or a thin film of oil upon water, change with the incidence of light and the angle from which they are viewed. A hummingbird's gorget may gleam intensely as he faces the watcher, to fade as suddenly when he turns sideward — a fact that we must remember when we consider his displays to a female or a rival. The duller feathers of these glittering birds, principally on their wings and tails, usually purplish black, brown, rufous, or cinnamon, are caused by pigments, mainly melanin, as in most birds. Hummingbirds lack red and yellow pigments.

One might suppose that to have a graceful, streamlined body, clothed in glittering contour plumage, would be sufficient adornment for such diminutive creatures. Nevertheless, many male hummingbirds, including some of the tiniest, are lavishly embellished with ornamental plumes, including long crests, projecting ear tufts, and long, colorful "beards." The tails of these more ornate hummingbirds are exceedingly diverse: slender, flexible streamers longer than their bodies, sometimes crossed; stiffer, graduated, elongate rectrices; incurved tail feathers; outcurved tail feathers; long rectrices tapering to sharp points; racketlike feathers with long, bare shafts terminated by broadly expanded vanes, like those of motmots but more elaborate. Several of these features may be combined on the same male. How can we account for this great diversity of ornaments which, far from contributing directly to the survival of their bearers, sometimes appear to impede them as they fly from flower to flower or dart erratically, catching tiny insects in the air?

The 320 species of this charming family — second largest of the avian families confined to the Western Hemisphere — are spread over the continents and islands from Alaska to Tierra del Fuego. They are most numerous in the equatorial zone, where they range from warm lowlands to as high on Andean peaks as flowers bloom on the verge of perennial snow. From this center of abundance, they diminish in number northward and southward. Those that nest where winter is severe migrate to avoid it.

As far as known, only in two closely related species, the Rufous-breasted and Bronzy hermits (*Glaucis hirsuta* and *G. aenea*), do male hummingbirds take an interest in eggs or young, and these do no more than guard the nests of one or two mates, occasionally inspecting them. All other male hummingbirds (except an occasional abnormal individual) limit their contribution to inseminating the females, whom they try hard to attract. To advertise their availability, male hummingbirds have two widespread methods, which we may call the dynamic and the static, or visual and vocal. They make spectacular flights, often rising high into the air, then shooting downward, to the accompaniment of vocal or mechanical sounds; or they gather in singing assemblies, where each, perching in his own small territory, proclaims his presence by more or less melodious notes. The presence of a number of performers in a traditional spot, occupied year after year if the vegetation remains appropriate, compensates for the weakness of their voices and helps the local population to become familiar with the situation of the gathering. This static method of attracting females is employed mainly by hummingbirds who live amid dense tropical vegetation that would obstruct extended aerial displays. The dynamic system is found chiefly among hummingbirds of more open country, near or above timberline on tropical mountains, and in the temperate zones at forest edges, in thin woodland, or in chaparral.

A singing assembly may be situated high in trees or near the ground. For obvious reasons, the former are more difficult to watch. The assemblies that have most invited careful observation are those of hermits, as the hummingbirds of subdued coloration that prefer the lower levels of heavy tropical forests, light woods, or thickets in humid regions are called. A well-studied species is the Little Hermit (*Phaethornis longue-marus*), only 3.5 inches (8.5 centimeters) long, widely spread over tropical America from southern Mexico to western Ecuador and Amazonian Brazil, and from the lowlands up, sparingly, to about five thousand feet (1,500 meters). Although in some regions it prefers the undergrowth of humid forests, in Central America it frequents forest edges, light second-growth woods, and tall thickets.

While making my way laboriously through a tangled growth of bushes and vines in light woods, I sometimes come to a spot where, all around me, small voices arise from unseen sources. The authors of the notes would be difficult to detect in the dimly lighted undergrowth if the rhythmic up-and-down wagging of their white-tipped tails did not draw attention to their tiny brown bodies, perching on thin twigs and vines, from a few inches to two feet above the brown fallen leaves with which they blend. Not shy, they permit me to approach within a few yards, with unavoidable noises as I advance through the obstructing growth, before one rises slowly, like a toy balloon released from the hand, to settle lightly on another twig at no great distance.

Separated from their nearest neighbors by about ten yards, the dozen or two Little Hermits who compose one of the larger singing assemblies may spread over fifty yards or more. Each is to be found on his own perch day after day through a long breeding season. Here in the Valley of El General, where the dry season is usually short, they fall silent in mid-February, when the drought is most severe and flowers fewest, to resume singing in April, when renewed rainfall has refreshed the vegetation. Likewise they sing little, chiefly in the early mornings, when rains are heaviest toward the end of the wet season, from September to November. With these exceptions, the hermits are present in their assemblies throughout the year.

Although hardly brilliant, the Little Hermit's song is longer and more complex than the monotonous squeaking of its larger relative, the Long-tailed Hermit (*P. superciliosus*). One version that I heard began with a measured *chip chip chip*, followed by a rapid, lilting *do da do a de* in a higher pitch. Sometimes the whole refrain consists of five notes, the first two delivered hurriedly, the final three more deliberately. Now and then the song sinks to a whisper. David W. Snow, and later R. Haven Wiley, tape-recorded Little Hermits' songs in Trinidad and demonstrated that close neighbors have similar songs, different from the songs of other groups in the same expanded assembly. Years earlier, I was puzzled by the great differences in the songs of White-eared Hummingbirds (*Hylocharis leucotis*) in different assemblies on the same Guatemalan moun-

Little Hermit, *Phaethornis longuemareus*. Sexes similar. Southern Mexico to western Ecuador and Amazonian Brazil.

tain. Now we have enough evidence from various sources to conclude that hummingbirds learn their songs, probably by elaborating upon an innate foundation, as in certain songbirds. The young hummingbird who joins an assembly learns to sing like his nearest neighbors. Some Little Hermits sing with a pleasing cadence, but in voices too thin to satisfy human ears.

At the height of the breeding season, each Little Hermit repeats his song throughout a long day, in bright or gloomy weather. In an all-day watch, David and Barbara Snow found that one male was present in his territory for a total of 444 minutes, or 70 percent of the 10.5 hours of their record. Performing at the rate of about thirty songs per minute nearly the whole time he was on his perch, this little bird delivered about twelve thousand songs in a day. On this day the hermit left his territory forty-eight times, for intervals ranging from less than two minutes to rarely more than eight minutes. These brief absences were probably spent foraging at a distance, for the singing assemblies of Little Hermits, like those of other hummingbirds, are not always situated where nectar-yielding flowers are abundant. These male hummingbirds defend territories not for the food they contain but as stations for attracting females.

As we have seen in the Black Grouse, Sage Grouse, Ruff, and Great Argus Pheasant, and shall presently see in the case of manakins, mating commonly occurs on the lek where the males display. In the more volatile hummingbirds, this is only exceptionally true. One can spend hours watching hummingbirds sing in their assemblies without seeing a single mating. A main difficulty is the impossibility of watching simultaneously more than one or two of the performing birds scattered amid obstructing foliage. To be sure, a watcher often sees one hummingbird closely approach another singing on his perch, but nearly always both fly swiftly beyond view before one can determine the sex of the visitor, especially in those many species with singing assemblies in which sexual differences in plumage are minor or lacking.

All of an August morning, I sat in a blind in a grove of tall second-growth trees in view of two members of a Little Hermits' singing assembly. Several times I saw one of these birds and another of unknown sex and provenance hover face to face on vibrating wings while they floated slowly upward, one sometimes slightly above the other. Suddenly, they darted away so swiftly that I could not tell who was the pursuer. Such is the usual outcome of these encounters. In a less frequently witnessed interaction at this assembly, a hermit hovered in front of and above another, who was perching about a foot above the ground. Floating nearly upright in the air, a little above the stationary one, it swung from side to side through an arc of one foot or less. Then it assumed a horizontal position, with head and spread tail bent strongly upward, giving the whole hummingbird a crescent shape, the horns pointing skyward. In this curious posture, it wafted from side to side, at the same time rotating alternately clockwise and counterclockwise through about 180 degrees to change the direction it faced. Or it might rotate in this manner at a fixed point in the air. Suddenly, performer and spectator darted away. The stationary member of this couple occupied the perch often used by the more persistent of the two singers in view of my blind; but whether this bird was displayed to by another hermit, or whether he displayed to one who had settled on his perch in his absence, I could not tell.

In the forest, on a morning at the end of March, I watched a more wonderful performance. One hermit, whom I took to be a female, rested upon a slender dead twig about a foot above the ground, while another, apparently male, hovered a few inches above her upturned head, on wings beating invisibly fast. With head and tail inclined strongly upward, he reminded me of a tiny boat with a high bow and stern, floating upon an invisible fluid. Making a sharp humming sound like a bumblebee, he oscillated gently back and forth over a distance of a few inches and, simultaneously, more slowly up and down. Every few seconds, he about-faced, and at longer intervals he revolved rapidly through a full circle, or even a circle and a half. He varied his act by shooting wildly back and forth for a foot or two, while his wings buzzed more loudly and insistently. After each of these interludes of more vigorous display, he resumed the quieter floating above the female, with gentle oscillations up and down and back and forth, and frequent rotations in the air. I was so enchanted by this charming display that for a good while it did not occur to me to look at my watch. After I started to time it, the performance continued for five minutes. I estimated that the hermit had displayed in my presence for a good ten minutes, and he was already so engaged when I found him. I admired his endurance.

To me, towering above the hovering hermit, his most conspicuous color was the chestnut of his rump and upper tail coverts. To the supposed female, viewing him from below, his cinnamon breast and whitish thighs, with feathers puffed out in prominent tufts, must have been his outstanding features. In contrast to certain hummingbirds with glittering crowns, who bend their heads downward while performing somewhat similar shuttle displays, the hermit kept his head elevated; on it was nothing special to show the female. During the whole performance, she held her long bill pointed straight up toward him; when he shot back and forth in longer arcs, she moved her head to follow. She appeared to be intensely interested in what he did. Finally, the tiny acrobat shot away through the underwood; then she for whom he had performed promptly vanished, too. This appeared to be a courtship ceremony; yet this was far from the site of the nearest singing assembly of which I knew, and now, at the end of a severe dry season, the assemblies were inactive.

One October, while cutting a trail through heavy forest near the Pacific coast of Costa Rica, I came upon a patch of large-leaved *Heliconia* in an opening amid the great trees, where a female rested upon a low twig, while a male displayed above her, much as I have already described. I do not know how long he had been engaged in this, but a minute or so after I arrived, he settled upon her back, whereupon she moved to avoid him and alighted a short distance away. After a renewed display above her, the male again tried to mount her. Again she flew, this time beyond my sight. Here, too, was no courtship assembly, at least at this season.

Although delightful to watch, the hermits' displays perplex the observer who tries to interpret their significance without being sure of the birds' sexes. In a version of the display that I witnessed in April, the roles of the participants were reversed. While one Little Hermit perched a few inches above the ground, swinging its tail rapidly up and down, another performed above it, as already described. After a while, the spectator rose into the air and the performer settled on the perch, to become the passive watcher of an act like that it had just given. This reversal was repeated several times, and I detected no difference in the aerial performances of the two individuals. Then both rose into the air, and one chased the other back and forth above the perch; or perhaps their roles were again reversed, and they chased each other alternately. Finally, they flew off through the woods together. This also occurred at a distance from a singing assembly.

My belief that Little Hermits display to females more often at a distance from the singing assemblies than in them is strengthened by observations made by T. A. W. Davis in British Guiana (now Guyana) on the Reddish Hermit (*Phaethornis ruber*). Even smaller than the Little Hermit, this tiny hummingbird of second growth and forest edges is widespread in tropical South America east of the Andes. The rufous-buff breast of the male is crossed by a black bar which is faint or absent in the female, thus making the sexes more readily distinguishable than those of the Little Hermit. At their gatherings in low, dense undergrowth, Reddish Hermits sing *zee zee zee zeezeze*, the song lasting about two seconds. While a female, away from an assembly, perches about five feet up, the male hovers in front of her, his body almost upright, his fanned tail pointing straight up, all his feathers puffed out, his long white tongue protruded so far that it appears to double the length of his bill and sometimes seems to touch the female's bill. While he hovers so, his wings droning loudly and deeply, he slowly revolves, thereby displaying every part of himself to the sitting bird. He was not seen to shuttle from side to side, in the manner of the Little Hermit. Toward the end of such a display, he repeatedly flicks his tail forward, perhaps striking his back with it, making a dull *whack* that rises above the continuing drone of his wings. Meanwhile, the watching female warbles softly and sweetly. As the male leaves at the termination of such a display, the female may pursue and induce him to repeat the act while she rests upon another low twig.

The several displays watched by Davis were at a distance from any known singing assembly. Barbara Snow witnessed similar displays in an assembly, but all were apparently incited by the visit of one male to another's territory. During one of these displays, she seemed to hear the perching male utter the soft, sweet *weep weep* in time with the movements of the hovering bird. While the latter performed, she noticed a peculiar feature that has also been reported of other hermits. Opening his mouth to display the interior, he spread apart the flexible basal branches of his lower mandible, thereby quadrupling the width of his bright yellow gape. Like the Little and other hermits, the Reddish fastens its downy nest with cobweb beneath the tapering tip of a leaf, usually a frond of a small, spiny palm, which forms a roof above it.

Much larger than the two preceding species, the Green Hermit (*Phaethornis guy*) ranges from Costa Rica and Panama through northern South America. Its singing assemblies, often attended by a dozen or more males, are usually situated in the undergrowth of rather heavy, humid forest. Each hermit perches no more than a yard or two above the ground, vigorously swinging his long, white-tipped tail up and down while, with clocklike regularity, he monotonously repeats a loud, barking monosyllable, which has been transcribed as *waatch*. In Trinidad, Barbara Snow counted from thirty-five to sixty-five of these notes per minute, with the rate rising to about eighty per minute in the excitement attending the visit of a female.

When about to alight on his singing perch, or when he finds a visitor resting there, the hermit hovers above it and darts from side to side for a foot or two, in a shuttle flight not unlike that of the Little Hermit. At the moment of reversing the direction of his swing, he opens his bright scarlet gape and emits a loud, explosive *tock* that sounds mechanical but is actually vocal. Light shining through the skin stretched between the

Green Hermit, *Phaethornis guy,* male singing.
Costa Rica to southeastern Peru.

spread basal branches of his lower mandible (as in the Reddish Hermit)
flashes brightly in the dark forest. Only one *tock* is sounded with each
reversal of direction, but the shuttle flight is so rapid that these explosive
notes follow in quick succession. After settling on his perch, the Green
Hermit utters a high-pitched *tsee,* and opens his uptilted bill to display
again his scarlet gape. When a visitor, male or female, approaches a male
on his singing perch, the two *tock*-display alternately, exchanging places

on the twig in a rhythmic sequence much like that which I watched two Little Hermits perform. Rising above the perch, one of the hermits swings from side to side emitting *tocks,* then displaces the resting bird, who floats up to give a similar performance. They may change roles a dozen times without interruption.

A female comes alone to the assembly, flying well above the males' perches and usually visiting several of them within a few minutes. Her arrival stimulates all the males to sing more rapidly and to wag their tails faster and more vigorously. As she hovers above the male of her choice, he stops calling and opens his mouth widely. With the tip of her bill, she often touches the scarlet lining, his most vivid color. Then, he usually rises and gives the *tock*-display just below the female, again showing her his brilliant gape. She may alight on the male's perch and exchange *tock*-displays with him. Twice Mrs. Snow saw a male mount a visiting female on his perch. Whether or not coition occurs, they fly off together, possibly to mate at a distance. With cobweb, the female Green Hermit fastens her nest beneath a narrow strip of a large leaf of *Heliconia,* which like the similar leaf of the banana plant readily tears from margin to midrib, or under the tapering tip of some other leaf.

Although it is technically correct to designate the Green Hermit's monotonous bark a "song," since functionally it corresponds to the advertising song of more melodious birds, this might seem to be a misuse of the word. The vocal outpourings of a few other hummingbirds are so charming that one does not hesitate to call them songs. One of these more gifted singers is the Band-tailed Barbthroat (*Threnetes ruckeri*), a hermit who, like the Green, proclaims his presence while perching near the ground in the dark undergrowth of wet forest. After singing alone for at least three years, a barbthroat in my forest was joined by a second, and, after three more years, by a third. They formed a very loose assembly, with the two farthest apart separated by about 250 feet (76 meters), but no others were found in this tract of woodland. Although the Band-tailed Barbthroat's voice is thin, his verse is varied and tuneful, one of the few hummingbirds' songs I have heard that seems to be delivered with feeling rather than mechanically. When I first heard a barbthroat amid tangled undergrowth, I looked for a wood warbler or some other songbird, and was not convinced of my error until I had laboriously traced the voice to its source. The song was plaintive, suggestive of melancholy, and often so rapid that it was almost a trill. The barbthroat's songs, lasting four or five seconds, are separated by much longer intervals of silence. Like other hermits, he incessantly waves his revealing tail up and down while he rests on his song perch.

All the foregoing hummingbirds live at low and middle altitudes. The best singer of all that I have heard prefers cool heights. On a steep, bushy slope at about nine thousand feet (2,750 meters) above sea level in Guatemala, four tiny male Wine-throated Hummingbirds (*Atthis ellioti*) sang on bare, exposed twigs, twenty-five to thirty yards (23 to 27 meters) apart. Whenever I passed that way, I stopped to listen, enchanted by a song that amazed me, coming from so small a throat. Although weak, the voice was not squeaky. For thirty or forty seconds without pause, a Wine-throated Hummingbird poured forth a song so intense, so varied with rising and falling cadences, that it reminded me of the higher notes of a small finch, such as the White-collared Seedeater (*Sporophila torqueola*) of the lowlands. While he sang, the hummingbird spread his gorget to form a scaly shield covering his throat, with the longer feathers at its sides projecting as sharp points at the lower corners, and turned his head from side to side. When he faced me, his gorget shone with intense magenta; as he slowly turned his head away, the color dimmed through metallic green to velvety black when viewed from the side. At times the little singer vibrated his wings and either floated slowly to another perch or hung in midair on invisible wings, all without interrupting his song. Or he flew in a long loop that brought him back to his point of departure, singing all the way.

Hummingbirds more brilliant than hermits often sing in their assemblies on higher, more exposed perches. One of the most familiar is the Rufous-tailed (*Amazilia tzacatl*), a glittering green bird which from Mexico to northwestern South America frequents gardens, plantations, and lightly wooded areas at low and middle altitudes. For twenty-two years, a singing assembly, which at different times had from two to four members, was established in our garden. Perching conspicuously from ten to twenty feet up on slender twigs, from fifty to seventy-five feet apart, these Rufous-tails began at dawn to repeat a quaint, subdued song that lacked the variety and animation of the vocal performances of certain other hummingbirds. *Tse we ts' we*, or *Tse we ts' we tse we* they tirelessly proclaimed until sunrise or a little later, then were mostly silent for the rest of the day. Often one of these birds interrupted his singing long enough to rise to a height of twenty to thirty feet above the ground and hover there for a few seconds before returning to his perch, in an aerial display much less spectacular than that of many hummingbirds who do not gather in assemblies. This singing assembly was active through most of the year, except toward the end of the dry season, in February and March, when flowers became rare. In April or early May, when showers had refreshed the earth, singing was resumed, to continue through the long rainy season, with some diminution of intensity during the wettest months.

Another singing assembly in our garden was composed of Beryl-crowned, or Charming, Hummingbirds (*Amazilia decora*). For a number of years, four males sang amid the shade trees, on perches mostly eight to twenty feet above the ground and fifty to one hundred feet apart. Their times of singing changed in an interesting fashion through most of the year. In certain months, such as the beginning and end of the singing season, they performed chiefly or only in the morning, from early dawn to about sunrise. As their ardor increased, they sang in the early morning and again for about an hour at the day's end, with sometimes a little sporadic singing between these two principal periods, especially on dull, cloudy days. But when their zeal was greatest, as in November and December when the wet season passed into the dry, we heard their sharp, metallic *tsweet tswe we we weee, tsweet tswe tswe we we we* . . . all day long. Like their neighbors, the Rufous-tails, they sang little at the height of the dry season in February and March. Other singing assemblies of Beryl-crowned Hummingbirds are in tall but light second-growth woods.

A third hummingbird that has sung in and around our garden is the large, rather plain Scaly-breasted (*Phaeochroa cuvierii*). As though to compensate for its lack of adornments, its songs, of which I have heard many versions, are often more complex and pleasing than those of its neighbors. One sounded like *cheee twe twe twe twe* — trill — *chup chup*. The trill, which first caught my ear, was so like the little trill of the Common Tody-Flycatcher (*Todirostrum cinereum*) that I looked for this diminutive yellow-breasted bird and was surprised to find a hummingbird. A frequent version runs, as well as I can paraphrase it, *see seea chweee, see seea chweee, sea sea chip-chip-chip-chip-chip*. The *chweee* is a low, full note, unusually strong and mellow for a hummingbird, and clearly audible at a distance of two hundred feet (60 meters). In addition to the foregoing verses with their many variants, Scaly-breasts sing a sort of medley, composed of the faintest notes, too slight to be called squeaky. Indeed, much of this whisper song is too weak to be audible to me at a distance of seven or eight yards, if not too high-pitched to be detected by the human ear at any distance. Nevertheless, after the tenuous medley dwindles into silence, I can be sure the bird is still singing by watching his distended, vibrating throat. Such subdued medleys are not rare among hummingbirds, especially juveniles learning to sing. They might be compared to the medleys of young songbirds, such as wrens, before they have acquired the more stereotyped, often less pleasing, adult song.

Scaly-breasts perform in their gatherings throughout the day, from early morning to late afternoon, and through most of the year, except in the driest months. The exposed twigs on which they sing are often in small trees that rise above a low, tangled thicket, or in the shade trees of a coffee plantation, usually from about thirty to forty feet above the ground. The two to four birds who commonly compose an assembly are separated by about one hundred feet or a little more. Between songs, they fan out their tail feathers, revealing the white tips of the outer ones, and vigorously wag their spread tails while they shake their relaxed wings.

In aggregate I have spent many hours watching the singing assemblies of these and a number of other species of hummingbirds that perform well above the ground, without ever seeing a nuptial display, such as I and others have witnessed among hermits that stay much lower. When a hummingbird singing on a high perch is visited by another, the two usually fly rapidly beyond view before one can distinguish their sexes. If the second bird is not an intruding male but a female, she apparently accompanies the resident male for courtship and mating in some more secluded spot. The paucity of flowers around many of these assemblies shows clearly that these hummingbirds do not sing to proclaim possession of feeding territories. The behavior of hummingbirds defending flowers is quite different from that in the assemblies where they sing to attract females.

The singing assembly is the prevailing mode among tropical hummingbirds whose courtship habits are known, but these are only a small minority of the hundreds of species, most of which remain to be investigated — and hummingbirds that sing in assemblies are more likely to attract attention than those that perform alone. In contrast to the situation in the tropics, singing assemblies are not known much beyond them. The eight species that breed north of the southern fringe of the United States have very different habits. Although all have sharp or squeaky calls, only the single permanently resident species, Anna's Hummingbird (*Calypte anna*) is credited with song. All eight of them depend upon flight displays rather than voice to make their presence known. Rising high in the air, they swing repeatedly back and forth, tracing a U with arms occasionally as much as a hundred feet high, or a more open semicircle up to twenty-five feet wide. As they descend at great speed in powered flight, air rushing through the modified feathers of wings or tail produces a shrill note, which in the case of Costa's Hummingbird (*Calypte costae*) has been compared to "the shriek of a glancing bullet, or a bit of shrapnel." At the nadir of the trajectory is often a perching female, or some other bird that the acrobat tries to expel from his territory. These high, alternating flights remind us of the back-and-forth shuttle flights that we noticed among hermits, magnified a hundredfold to take advantage of the more open spaces where these northern birds live. They would obviously be impossible in tropical forests and thickets.

In all eight of these northern hummingbirds — Ruby-throated (*Archilochus colubris*), Black-chinned (*A. alexandri*), Anna's, Costa's, Calliope (*Stellula calliope*), Broad-tailed (*Selasphorus platycercus*), Rufous (*S. rufus*), and Allen's (*S. sasin*) — the males have scintillating metallic gorgets, some shade of red, blue, or purple, which they flash in the face of the target of their aerial displays as they shoot close above it. In all, the female lacks the gorget and is much plainer than the male. In hummingbirds that sing in assemblies, sexual differences in coloration tend to be slighter; often the sexes are difficult to distinguish. Another hummingbird with a high flight display is the Lucifer (*Calothorax lucifer*), which from the Mexican highlands reaches the extreme south of the United States in Arizona and western Texas, in open, mostly arid, country. The male wears an elongated violet gorget, which the much less intensely colored female lacks.

As an example of hummingbirds with high aerial displays let us take Anna's, probably the most studied of all hummingbirds because it lives throughout the year in California, where there are many professional ornithologists and amateur bird watchers. Gary Stiles tells that during the breeding season, which in California's Mediterranean climate with winter rains extends from about November or December to April or early May, males occupy territories amid chaparral, especially where the two species of gooseberries, *Ribes malvaceum* and *R. speciosum*, that are its principal sources of nectar grow and flower abundantly.

Each solitary male claims a core area of about a quarter of an acre (0.1 hectare) surrounded by a buffer zone of five to ten acres (2 to 4 hectares) that he uses less frequently and defends less consistently. Here he advertises his presence by a song that to Stiles sounded like *bzz-bzz-bzz chur-ZWEE dzi! dzi! bzz-bzz-bzz*. He also performs the most elaborate and spectacular dive displays of any hummingbird in the United States. Each begins while he hovers about six to twelve feet above the object of his display and delivers one or two buzzy songs. Then he flies in a wavering course almost straight upward to a height of fifty to one hundred feet. From this high point he dives almost straight downward, to level off and shoot at great speed two or three feet above the display object, with an explosive squeak as he passes closest to it. Without pausing, he loops upward to his starting point, where he sings again before rising for another powered dive. He may give from five to ten of these breathtaking displays without resting. If the sun shines brightly, he flies directly toward it as he passes over the display object, so that its reflected rays make the brilliant rose-red of his gorget and crown flash more dazzlingly in the viewer's eyes. The target of his display may be a male Anna's Hummingbird or some other small bird invading his territory, an intruding human, or a visiting female of his kind.

Although Stiles never followed a courtship sequence from start to finish, he repeatedly witnessed its components, and by piecing them together formed a picture of its probable course. A female building a nest flies to a male's territory and visits his gooseberry flowers. He detects her and approaches with song or chatter. She perches and becomes the target of his dive displays. When she flies toward her nesting territory in a different habitat, he follows. If she rests along the way, he dive-displays to her. When she shifts her perch, he follows, to sing vehemently to her, or to shuttle-display a foot or less above her, swinging rapidly back and forth through an arc of six to ten inches (15 to 25 centimeters), reversing direction with a flick of his spread tail, and holding his head bent down to show his glittering gorget and crown (not upward, like the oscillating Little Hermit, who has no similar elegance to display). While shuttling back and forth, the male repeats *bzz* notes, and at the end of this exhibition he may perch nearby and sing with high intensity. While the courting male swings back and forth above her, the female follows his movements with her uplifted bill — whether fascinated by his glitter or to hold him, so to speak, at sword's point is not clear. If sufficiently impressed, she permits this indefatigable suitor to mount her. Then he flies back to his territory, while she resumes work on her nest, where she will never receive his help.

The intimate shuttle flight of the courting male has been largely overlooked but has now been recorded for all eight northern hummingbirds as well as a number of species in Mexico and farther south in tropical America, including hummingbirds that sing in assemblies and those with high aerial displays. In some species it is vertical instead of horizontal, but always it is performed close to a perching, or sometimes hovering, female. It appears to be the most important element of courtship, as it permits her to view closely whatever adornments her suitor may have, as swift display flights fail to do. It is a pity that we lack adequate accounts of any of the more lavishly attired tropical hummingbirds, such as those with racket tails, long streamers, or high crests, to know whether shuttle displays are included in their repertoires. Although I have never followed the complete courtship of the tiny, lavishly ornate White-crested Coquette (*Lophornis adorabilis*), I have on several occasions watched a shuttle display. While a female rested upon a low, slender twig, a male oscillated from side to side in front of her in a most peculiar lateral flight, always keeping his breast toward her. Swinging now toward his left wing, now toward his right, he changed direction with surprising suddenness. The perching female followed his movements with her head, always pointing her bill directly toward him. When she rose slowly above her perch, hovering on wing as only hummingbirds can, he continued to float before her, now oscillating more slowly.

At high altitudes in the tropics, as in the North Temperate Zone, certain male hummingbirds engage in spectacular aerial displays. In Guatemala I watched Broad-tailed Hummingbirds trace towering U's above a deforested slope at nearly ten thousand feet above sea level. Jean Dorst has described how that persistent singer, the Sparkling Violet-ear (*Colibri coruscans*), rises by stages as much as a hundred yards above the Peruvian puna, sings at the zenith of his trajectory, then dives headlong down to a low shrub. He traces diverse courses in the air, sometimes closely accompanied by a female flying parallel to him.

On the same high plateau, the sexes of the Andean Hillstar (*Oreotrochilus estella*) occupy separate territories, the males on the more exposed puna, the females in sheltered dells with rocky walls to which they attach their nests. The courtship of this rather large hummingbird proceeds in three stages. The first is acted in the territory of the male, who ascends high above a visiting female perching low, then shoots downward with a vibratory sound made by air passing through his outer tail feathers. He continues his aerobatics, tracing more or less complicated arabesques in the air. In the second phase, the female flies close beside and parallel to him in similar flights, which may take the pair to the female's territory, often close to the nest she is building. Here, in the third phase, he perches, puffing out his plumage until the widely separated feathers of his metallic green gorget appear black. While she flies away, he continues to rest, nervously preening himself. She returns, alights beside him; he opens his mouth, exposing the vivid yellow interior; she pushes in her bill and feeds him, as though he were a fledgling. Probably this makes him feel more at ease on a territory from which, under other circumstances, he would be expelled. This reversed courtship, rare among birds and apparently without parallel among hum-

Andean Hillstar, *Oreotrochilus estella,* male. High Andes from Ecuador to
northern Chile and Argentina.

mingbirds, may continue for about half an hour, with repeated feedings
of the male by the female, while the two manifest great agitation that
may intensify to a brief squabble. All this culminates in coition, which
occurs immediately after the female feeds her partner, and may be repeated several times. Finally, the male returns to his territory, terminat-
ing the transitory association of the sexes.

In all these diverse courtship patterns of hummingbirds, the female
freely chooses her partner. Unable to compel her, he must make himself
attractive to her by his appearance, his voice, or his displays, singly or
combined. Female preference has promoted the splendor of these exceptionally refulgent little birds.

References

Bent 1940; Carpenter 1976; T. A. W. Davis 1958; Dorst 1956, 1962; Hamilton 1965; Skutch 1964a; 1964b; 1967; 1972; 1973; 1981, 1983b; B. K. Snow 1973a; 1973b; 1974a; D. W. Snow 1968; D. W. Snow and B. K. Snow 1973; Stiles 1973; 1982; Stiles and Wolf 1979; Wagner 1946; 1954; Wiley 1971.

8.

The Courtship of Manakins

Although not closely related, manakins and hummingbirds have much in common. Small to very small, they are among the most abundant birds of tropical American woodlands. Both families contain outstandingly beautiful species, but their brightest colors have different sources. The first view of a male hummingbird is often disappointing; his metallic brilliance is caused by the minute structure of his feathers; unless the light and the angle of vision are just right, he may appear disenchantingly dull. It is otherwise with male manakins, whose vivid colors, due to pigments, delight the eye from almost any angle. When I first saw Long-tailed Manakins (*Chiroxiphia linearis*) amid dense undergrowth of a Guatemalan woods, I thought of birds of paradise. The first glimpse of a Red-capped Manakin (*Pipra mentalis*) rarely fails to evoke exclamations of delight from bird watchers. Not only by their appearance but also by the unexpected sounds they make and by antics peculiar to themselves do manakins win attention. The fifty-three species in this family spread over continental America from southern Mexico to northern Argentina, where they prefer warm lowlands. They are absent from the West Indies, except Trinidad and Tobago.

Both manakins and hummingbirds gather much of their food in flight. Hummingbirds suck nectar from flowers and catch minute insects in the air while hovering in their inimitable way; manakins pluck berries from trees and shrubs by darting up to them without alighting, and they glean insects from foliage in much the same manner. In contrast to insects, which try to avoid capture by concealment, nectar and fruits are freely offered by plants, which advertise their availability to animals that pollinate their flowers or disseminate their seeds. The ease with which these foods can be collected in the seasons of their abundance permits manakins and hummingbirds to adopt life styles rare among birds more dependent upon insects. In small, open nests, females of both of these families lay two eggs and, without male assistance, raise their two young, which is also the number most frequent in nests of small tropical birds attended by both parents. Relieved of domestic chores, male manakins and hummingbirds can devote their energies to courtship, which many prefer to do in close association with others of their species and sex. They evolve bright colors and adornments, while the females tend to remain much plainer, less likely to draw attention to their nests. Although in some species of hummingbirds females differ little from the glittering males, all female manakins wear cryptic plumage, usually greenish or grayish olive.

With the exception of the large, dull-colored, melodious Thrushlike Manakin (*Schiffornis turdinus*), which possibly is more closely related to the cotingas, manakins prefer to court at least within hearing of others of their kind. Male manakins exhibit three degrees of sociability: (1) They gather in courtship assemblies, or leks, where each performs alone in his territory. (2) Several males perform in the same spot (such as a fallen log) without coordinating their movements. (3) Two or more males perform highly coordinated displays on the same site.

Terrestrial Court Displays of Collared Manakins

As an example of manakins with grouped individual territories (as in the Black Grouse, Ruff, and many hummingbirds) let us take the genus *Manacus*. The males of all four species have the same basic pattern, diversely colored. The White-bearded Manakin (*M. manacus*), widespread in South America, is black on the whole top of his head, mantle, wings, and tail. His cheeks, broad collar encircling his neck, and underparts are white. The White-collared Manakin (*M. candei*), which ranges through the Caribbean lowlands from southern Mexico to Costa Rica, is rather similar, but his posterior underparts are bright yellow, and he has a white wing-bar. The Golden-collared Manakin (*M. vitellinus*), confined to Panama and Colombia, has a yellow collar and olive posterior underparts. The Orange-collared Manakin (*M. aurantiacus*) of southern Pacific Costa Rica and adjacent Panama has a bright orange collar extending to the · breast, pale yellow belly, and olive instead of black tail. This bold pattern is absent from the females. Both sexes of the four species have bright orange legs, which distinguish the females from similarly olive or olive-green females of other genera.

Orange-collared Manakin, *Manacus aurantiacus,* male. Pacific slope of
Costa Rica and western Panama.

    The close relationship of the four species of *Manacus,* suggested by
the pattern of their plumage, is corroborated by their courtship. All perform at bare patches of ground, their "courts," from which they pick up
in their bills and carry aside all fallen leaves and other litter not too
heavy, until the area appears to have been swept clean with a broom.
These courts are situated, according to the species and the region, beneath old forest, often near its edge or beside an opening, in light second-
growth woods, or beneath tall thickets, usually amid thin sapling trees,
two or more of which stand upright at or a little beyond the edge of each
court. These bare areas are usually roughly circular or oval and vary in
size. Mostly they range from about twenty-four to thirty inches (60 to
75 centimeters) in diameter, but I have seen occupied courts of Orange-
collared Manakins only eleven by eleven inches (28 by 28 centimeters).
At the other extreme, a White-collared Manakin's court measured fifty-

four by forty-five inches (137 by 114 centimeters), and several others were about four feet (1.2 meters) long and nearly as broad. These large courts were surrounded by five to eight saplings. The young trees, indispensable for the manakins' displays, must be vertical and slender enough to be grasped by the birds' small feet, usually from a quarter to half an inch (6 to 13 millimeters) in diameter.

Except possibly where manakins are rare, their courts are not scattered at random through suitable vegetation but aggregated in assemblies. The courts on Barro Colorado Island where Frank M. Chapman made his classic study of Golden-collared (Gould's) Manakins were in groups of four to seven, with the individual courts twelve to two hundred feet (3.7 to 61 meters) apart, mostly thirty to forty feet (9 to 12 meters). The closest assemblies that he found were separated by about three hundred yards (274 meters). The largest lek of Orange-collared Manakins that I have seen contained fourteen occupied courts, scattered through about half an acre (0.2 hectare) of tall but light second-growth woods. These courts were 8 to 102 feet (2.4 to 31 meters) from their nearest neighbors. The two closest courts were separated by dense vegetation that screened their occupants from each other. A small assembly of White-collared Manakins consisted of three courts twenty-five to forty feet (7.6 to 12 meters) apart. In all these assemblies, the manakins on their courts can probably see, often imperfectly because of interfering foliage, only their nearest neighbors, but are probably within hearing of the louder sounds made by all of them.

On Trinidad, as on other islands with a relatively small number of species of birds, some species are amazingly abundant. Here, at one of the assemblies where David W. Snow made his prolonged, thorough study, about seventy courts were crowded into a space of approximately twenty by ten yards (18 by 9 meters). Most were only a few feet apart, some almost in contact.

So long as the vegetation remains suitable, these assemblies persist in the same place year after year; and often the same individual court is renovated, by removal of accumulated litter, for use in successive breeding seasons. Accordingly, at least the older female manakins in the locality should know just where to go when their eggs are ready to be fertilized. To attract them to an assembly, and to help young, inexperienced females to find it, the males produce a surprising variety of notes. While wandering through the woods, one with ears attuned to natural sounds is often guided to a lek by a noise like the snapping of dry twigs or the detonation of small firecrackers, mingled with sharp calls. If, after examining the bare patches of ground, and perhaps wondering what made them, one stands apart to watch, one may presently see the manakins, scattered by the intrusion, return to perform in a manner unique among birds.

Each little collared acrobat goes to his own court, where he jumps rapidly between saplings on opposite sides, tracing a low arc about a foot above the bare ground. If he has more than two saplings, he may jump around his court. As he shoots across his court, he makes the loud *snap* that drew attention to him. In a maneuver too swift to be followed by the human eye, he alights on each upright stem facing the direction from which he came, ready for the reverse leap, which follows immediately. He may jump only twice, or a dozen or more times in swift succession. Often he ends a series of leaps by descending to his bare court, where he may delay for two or three seconds before he springs upward, to the height of a foot or less, to the accompaniment of a harsh *grrrt*, or a sound between a grunt and a whir, according to the species. The performance of one manakin stimulates his neighbors to do likewise, until, in a large assembly, a chorus of snaps issues from the undergrowth.

During pauses between jumps, the manakin, clinging to one of his vertical stems, stretches his neck forward and protrudes the elongated feathers of his chin beyond the tip of his bill. In the two white-collared species, the "beard" is especially well developed. With the spread feathers of the collar, it forms a gleaming white border around the bird's black cap. Similarly, the Golden-collared Manakin surrounds his black cap with a yellow border. With extended beard, the manakin sometimes shakes his head from side to side.

The jumping display is too exhausting to be continued for long periods. Although, at the height of the season of courtship, a manakin may spend up to 90 percent of the day at his court, leaving it only for a few minutes at a stretch to forage, much of the time he perches near it instead of upon it, producing a number of characteristic sounds. All four species of *Manacus* frequently voice a clear, high-pitched call that has been variously written *chee-yú, pée you,* and *pee-yuh. Chee-pooh* often announces the beginning of a bout of display. A thin, tense *chee* expresses frustration or annoyance. All these notes vary subtly with the occasion or the bird's mood. The usually silent female voices a weaker *cheeu* when perturbed, especially when anxious for her nest or young.

In addition to their rather limited vocabulary, these manakins make a variety of mechanical sounds. The sharp *snap* that accompanies each leap across the court has already been mentioned. While perching near his court, the manakin leans forward, lifts his wings well above his back, and vibrates them into a blur, thereby producing a very rapid series of sharp, crackling notes, the rolling snap, which may be imitated by holding a thin, flexible strip of wood against the teeth of a rapidly revolving cogwheel. Frequently this volley is followed by a loud *chee-yú*. This sequence of sounds often stimulates neighboring males to repeat it. It may prelude a bout of jumping over the court. The ordinary flight of a male *Manacus* is accompanied by a fairly loud rustling or whirring sound. As he approaches his court, takes short flights in its vicinity, or leaves it, this *whirr* becomes fuller and deeper. On flights of a few yards or more, it is heard intermittently, while the bird traces an undulatory trajectory instead of his usual straight course.

These nonvocal sounds are most probably made by the birds' wings, although their movements are so swift that the details have not been elucidated, and their role in sound-production has been questioned. The secondaries, with unusually thick shafts and very stiff outer webs, appear to produce the staccato *snap* and the rolling *snaps*. The stiff outer primaries, with very narrow outer webs, evidently make the *whirr* heard in flight. Females and young males fly with at most a low rustle, and molting adult males who have lost some of their outer primaries fly less noisily than those in full plumage.

The darting back and forth of these strikingly patterned little birds and the medley of sounds vocal and mechanical that arise from a well-attended court at the peak of activity early in the morning make it conspicuous to the eye and ear. As though to increase the visibility of his displays, or to make it harder for a predator to approach unseen, the manakin sometimes concludes a dance by plucking a small living leaf from a low plant beside his court and carrying it away, or by tugging at a larger one that he cannot detach. However, he does not defoliate his surroundings as thoroughly as certain other manakins that do not clear the ground. More often, he drops down to pick a fallen leaf from the bare area, or without alighting he snatches it up and drops it a short distance away. If you deposit a few leaves or flowers on an active court and watch at a distance, you may see the owner approach and carry them off.

Each court belongs to a single adult *Manacus* and is respected by his neighbors, who rarely interfere. In crowded courtship assemblies in Trinidad, White-bearded Manakins prefer central sites, and in the absence of a central court's owner, males less favorably situated may display there, to abscond as soon as he reappears. Here Snow noticed occasional joint occupancy by two males, which was usually brief, but in one case continued for seven months. However, when both were present, only one displayed; the subordinate member of the pair performed only in the other's absence. Rarely, a bolder male tried to oust the owner, who chased the intruder around and around the display area. Or the two clutched and rolled together on the ground. This seems most unusual; I have never seen manakins of any species fight, nor read any other report of such violence. To learn what a Golden-collared Manakin would do if another violated his privacy, Chapman set a stuffed skin of a male on a court, where it was vigorously attacked by the proprietor, showing that manakins will fight if sufficiently provoked. Normally, they abide so strictly by the conventions of their courtship assemblies, as in an old and stable culture, that clashes are avoided. Although I have often seen manakins of a number of species chase each other, they never made contact.

In intervals of inactivity, owners of two neighboring courts often rest close together, but not in contact, upon some intermediate twig. At assemblies of Orange-collared Manakins, I noticed that these visits occurred, not midway between the two courts, but closer to the smaller of them, the owner of which appeared to be less active than his visitor. Sitting a foot or so apart, the males often puffed out their feathers, making themselves appear quite roly-poly. Sometimes they twitched their folded wings and quivered their whole bodies, continuing this for about a minute. Chapman received the impression that the more submissive of two Golden-collared Manakins resting close together "courted" his dominant companion. Although during the breeding season two adult male manakins appear never to display together on a court, they may do so in the off-season, when courts are only occasionally visited. Sometimes, while one wanders through light woods before or after the breeding season, a sudden outburst of snapping and *chee-yú* calls draws attention to a small party of manakins, including both adult and green immature males, jumping back and forth together, frequently high above the ground.

Until about a year of age, males wear greenish or olive plumage so similar to that of adult females that they can be distinguished only by behavior. During a long vigil at a display ground, a watcher's attention may languish, for, except in the early morning, activity is far from continuous. A greenish bird's arrival at a court alerts the watcher, often only to be disappointed. The adult male proprietor may ignore the visitor, or fly away leaving him there. Apparently, the manakins can tell the sex of an individual in female plumage more readily than ornithologists can.

When, finally, a female arrives, the assembly perks up. Alighting upon an upright sapling beside a male's court, she joins him in an animated performance. They jump back and forth, crossing each other above the bare patch, each alighting on the upright sapling that the other has just left, all to the accompaniment of the male's loud snaps. Sometimes it looks as though they chase each other back and forth, but the action is so rapid that it is difficult to tell who chases whom. Meanwhile, all the other males, stirred by this activity, are jumping and snapping, each trying to attract the visitor to his own place. After a brief intermission, the same two birds may dance together again. Or, after a few leaps over the court, the female may suddenly depart, perhaps to perform in sequence with one or more different males. But if her partner pleases her, she remains clinging to one of the saplings, while he drops to the bare ground, springs up with a *grrrt* to a position above her on the stem, then slides downward to alight on her back and fulfill the purpose of so much elaborate display. After insemination, the female goes off to finish her nest, lay her two eggs, and rear her nestlings with no help from their father.

A courtship assembly like that of *Manacus* offers an ideal situation for the operation of intersexual selection, yet it is difficult to detect the grounds for a female's choice of a mate. Adult males all look too much alike for us to distinguish them without banding, and their performances differ little. Nevertheless, some are preferred above others, perhaps because they perform more frequently, are older, or have cleared larger courts, which may indicate greater energy. Although now the plumage and courtship behavior of each species appear to be uniform and stable, perhaps in the distant past it was more labile, and female choice has stabilized the attractive patterns of their plumage and their unique displays.

Mossy Log Displays of White-throated and Golden-winged Manakins

Soon after sunrise in early April, I paused in my walk along a woodland path in the mountains of southern Costa Rica to watch White-ruffed Manakins (*Corapipo leucorrhoa*) flitting about ahead of me. While one chased another, a third gave a charming display such as I had never seen before. With his diminutive body nearly upright, plumage all puffed out, tail raised until it almost touched his back, wings beating slowly through wide arcs, he traced a strongly undulating course across the path as he descended toward a mossy log. Flying so slowly that he seemed barely to avoid stalling, he reminded me of a tiny black balloon with a gleaming white patch, his widely spread throat feathers, on its forward side. After alighting on the log with his ruff still fully expanded, he lowered his foreparts and bent down his head, as though attentively examining some minute object amid the moss.

In the surrounding humid forest, about four thousand feet (1,200 meters) above sea level, I found three more moss-covered logs where White-ruffed Manakins displayed. Ranging in diameter from about five to eighteen inches (13 to 46 centimeters) they were horizontal or slightly sloping and, with one exception, in spots with rather open undergrowth that did not obstruct the display flights. The three logs that I watched most carefully were frequented by three, three, and four adult males and one or more males in transitional plumage. Although I have rarely seen males of *Manacus* in transitional plumage, White-ruffs take longer to attain adult attire, probably about two years, and males intermediate between the grayish olive-green of females and juveniles and the glossy blue-black of adult males are seen at all seasons.

Male White-ruffed Manakins spend much time perching on slender branches of small trees near their display logs, usually from fifteen to thirty feet (4.5 to 9 meters) up. With feathers fluffed, adult and transitional males rest a few feet or yards apart, appearing to enjoy each other's company. Occasionally they preen. From time to time, one beats his wings to make a resonant *flap,* such as one hears when a manakin flies swiftly to a log; but White-ruffs lack the sharp *snap* of *Manacus* and some species of *Pipra.* Although most of the time the several males who share a log appear to be friendly, occasionally one chases another around and around through the neighboring undergrowth, voicing slight, shrill notes such as one hears from parties of both sexes foraging through the woodland. I never saw the pursuer catch the pursued, or any fighting.

Descents to a log were of two kinds. In addition to the slow, silent, bouncing flight that first claimed my attention, manakins frequently flew down so suddenly and swiftly that I was unaware of their approach until they were nearly upon it. Possibly these rapid descents began above the treetops, for others have seen White-ruffed Manakins flying over the forest canopy. Usually my attention was drawn to the oblique downward dart by a *flap* such as may be roughly imitated by jerking taut a piece of stout cloth held loosely between the hands. This sound was often followed immediately by sharp, harsh little notes, the full sequence becoming *flap chee waaa.* Occasionally, the noisy descent, instead of starting at a point beyond view, began as a manakin, approaching with slow undulations, abruptly accelerated his flight and, with a swift, jerky movement, made the *flap cheee waaa.* Sometimes a manakin alighted in this boisterous fashion beside another resting on the log, who, not surprisingly, was startled into flight. Or the displaying manakin flapped as he flew close above another on the log, to continue onward and alight in the bushes beyond it.

After descending to the mossy log, by whatever means, the manakin usually stood with ruff widely spread, foreparts depressed, and sometimes with his head turned sideways, as though scrutinizing the log with one eye. While standing so, sometimes for several seconds, he often twitched his folded wings. Often he hopped along the log, or flitted from one part to another. When several were on the log together, they did not interfere with each other; their movements were always quite independent, with neither coordination nor aggression — unless the boisterous descent of one beside another was a restrained threat. After a short visit to the log, a manakin left with the same spectacular, bouncing flight that usually took him to it. After flitting through the undergrowth, he might repeat the whole performance.

Young males, who differed from females chiefly by their whitish throats and perhaps spots of black on their plumage, displayed much as adults did, approaching the log with either the butterfly-like undulatory flight or the swift noisy descent. Standing on the log, either alone or with adults, they spread their smaller ruffs. The older manakins paid no attention to these yearlings. In many hours of watching, I never saw an undoubted female at the logs.

The White-ruffed Manakin ranges from Honduras to Colombia and northwestern Venezuela. Farther east, in Venezuela, Guyana, Surinam, French Guiana, and adjacent Brazil, it is replaced by the White-throated Manakin (*Corapipo gutturalis*), a rather similar bird with a white bib extending in a V on the breast instead of an expandable white gorget. In the Brownsberg Nature Reserve in Surinam, Richard Owen Prum found White-throated Manakins displaying socially on old, mossy logs lying in the forest. In an area 164 feet (50 meters) in diameter, he located seven of these logs, and 1,150 feet (350 meters) away was another group of logs frequented by the manakins. Some males appeared to be fairly constant attendants at the same site, whereas others moved from one group of logs to a more distant group.

Most often the White-throated Manakins approached a log by flying normally from a perch a few yards high and about ten yards away. Occasionally, they flew to it with slow undulations, somewhat in the manner of White-ruffed Manakins, their bodies nearly vertical, their wings fluttering rapidly. For a more dashing approach, they rose almost straight upward to a height of ten to twenty-five yards above the treetops, where with increasing intensity they called *seee, seee, seee,* then plunged steeply downward.

As a manakin neared a log, he voiced from two to ten high, thin *seee* notes, then stalled in flight and, with a single flash of his white wing-patches, made a muffled *pop* as he alighted. Immediately upon landing, he rebounded with a flap of his wings, uttering a sharp, squeaky *tickee-yeah*, to come down again a foot or so away, facing the point where he first alighted. On the log, he displayed by pointing his bill straight upward, prominently exposing the white patch on his throat and chest. Or, with hunched back and lowered head, he quivered his wings, revealing their white areas. Groups of up to seven adult and immature males performed together, with no attempt to coordinate their movements. On the contrary, they competed for control of the log, as I never saw White-ruffed Manakins do. After displaying for five to ten minutes at one log, several males might move together to nearby logs, sometimes as many as five in succession, in a single period of activity.

Prum was no more fortunate in seeing the culmination of the White-throated Manakins' displays than I was at logs of White-ruffed Manakins. Many years earlier, in what was then British Guiana, T. A. W. Davis had better luck. While he watched a party of between six and a dozen White-throated Manakins of both sexes flitting around and chasing one another through the trees, a female dropped down to an old, moss-covered, fallen trunk, where she was joined by a male who mounted her without any preliminary ceremony. Later, a female came to this log, where a male, crouching with wings fully spread horizontally, crawled slowly and laboriously toward her. Before he could reach the waiting female, another male chased him away, and she also flew up to a branch. This observation suggests that the mossy logs of *Corapipo* are mating stations as well as centers for the males' displays.

Another species that displays on fallen logs, or sometimes on exposed prop roots of great trees, is the Golden-winged Manakin (*Masius chrysopterus*), which is found at low middle altitudes in forests on both slopes of the Andes from northwestern Venezuela to northern Peru. The velvety black male has a bright yellow forehead and forecrown, a patch of red, orange, or brown on the hindhead, a yellow throat, and on wings and tail much yellow that is mostly hidden when they are folded. At the posterior edge of the yellow crown patch are two tufts of black plumes that can be erected as short horns on either side of the head. The female is olive-green and yellowish. In western Ecuador, Richard Prum and Anne Johnson found adult males defending territories about 80 to 130 feet (25 to 40 meters) in diameter, each with two to four logs or exposed roots. These territories were grouped, with two or three close enough together for their occupants to hear one another.

A displaying male Golden-wing flies down to a log from a nearby perch. Immediately upon alighting, he bounces up into the air to descend to a neighboring point on the log, facing the direction from which he came, much as White-throated Manakins do. Next, he bends forward until his bill nearly touches the log while the yellow feathers of his fore-crown and his black horns are erected, his body plumage sleeked down. Then, fluffing out his feathers, he bows rhythmically right and left, so low that he almost touches the log with his bill, often continuing this for the better part of a minute. Rarely, he stamps his feet on the log, rapidly but briefly.

Noteworthy are the synchronized performances by two males, either two adults or an adult and an immature bird. The participants may be the resident male and an immature visitor or two intruders while the owner is absent. Facing each other on the log, they bow simultaneously from side to side. A resident male and an immature bird repeatedly engaged in a more elaborate exercise, which began with a typical log-approach display. While one participant (A) flew down to the log, the other (B) waited at the spot where he would alight after the usual re-bound. Before A could land there, B leapt to the point where A first touched down. Then, crossing each other in the air, they bounced alter-nately back and forth, in a dance reminiscent of that of two Orange-collared Manakins leaping between vertical saplings on opposite sides of a bare court. All these performances were accompanied by much calling. Some were addressed to visitors to the logs who appeared to be females, over whom the resident male passed from side to side on his rebounds after landing, but in no instance did the courtship reach its climax.

Shuttle Display of the Pin-tailed Manakin

Another strikingly attired manakin is the Pin-tailed (*Ilicura militaris*) of the forests of southeastern Brazil, where it was studied by Helmut Sick and, in more detail, by Barbara and David Snow. A velvety black crown and mantle separate the male manakin's crimson forehead from his scar-let lower back and rump. His tail is black, with the two thin central feathers projecting about an inch beyond the others. The sides of his head and underparts are white or pale gray. The female is olive-green above, gray below, with only slightly projecting central tail feathers.

In November and December, the Snows found Pin-tails displaying in groups of two or, more often, three, each on a private territory about sixty-five to one hundred feet (20 to 30 meters) across, and within hear-

Pin-tailed Manakin, *Ilicura militaris,* male. Southeastern Brazil.

ing of his nearest neighbors. Here he sang and displayed on the branches of understory trees, as well as on his nuptial perch, a yard-long length of a horizontal limb about two inches (5 centimeters) thick. Early in the morning, the owner of this perch pulled at moss and tore at leaves above it, removing obstructions to his principal displays. Throughout the day, but most frequently in the early morning, each male Pin-tail, perching with the scarlet feathers of his rump conspicuously fluffed, repeated a song of five to eight notes, occasionally many more, all with a falling inflection that imparted a plaintive quality. He flew between his perches with a whirring sound, undoubtedly made by his wings. At intervals, he made a single loud *snap* as he traced a curving course from one branch to another about a yard away.

The Pin-tail's most characteristic display was nearly always made on his nuptial perch. To begin this act, he rested facing inward at one end of the cleared length of this branch, with his plumage compressed, his foreparts lowered to it, and his tail elevated. Then, taking off with a *snap,* he would fly in a low arc to the other end of the branch, where he alighted facing the way from which he had come. With another *snap,* he retraced his course to his starting point, where he again landed facing inward. In forty-four hours of watching, the Snows witnessed mating only once. When a female arrived in a male's territory, he jumped to the accompaniment of a clicking sound between his song perches, then flew noisily to his mating perch, followed by her. While she watched, he jumped a dozen times to twigs just above, clicking each time he leapt.

Then he flew back and forth close above her, in the shuttling, snapping display already described. Finally, after alighting on his starting point, he promptly rose in a high, semicircular course that brought him down directly upon the crouching female's back. After coition, he left the nuptial perch and started to sing. She preened a little before she flew away.

The Pin-tailed Manakin's display has features in common with those of *Corapipo* and *Masius*, with the difference that it is performed on a high horizontal branch instead of a prostrate log. With the difference that he uses horizontal instead of vertical perches, he jumps with a *snap* much like that of *Manacus;* yet he lacks the highly modified secondary wing feathers to which the loud snapping sounds of the latter are attributed. Moreover, although males of both *Ilicura* and *Manacus* fly with a whirring sound, the outer primaries of the former have unusually broad ends, whereas the corresponding feathers of *Manacus* are thin and stiff. Apparently, we still have much to learn about the exact manner in which these various sounds are produced.

Aerial Displays of Pipra

Pipra, with sixteen species the largest genus of manakins, exhibits the transition from displays by single males to coordinated performances by two, without, however, attaining the complexity of the elaborate dances of *Chiroxiphia,* the blue-backed manakins. As an example of manakins that display alone in courtship assemblies, let us take the Red-capped, or Yellow-thighed, Manakin (*Pipra mentalis*), a bird barely four inches (10 centimeters) long, that ranges through the more humid forests from southern Mexico to northwestern Ecuador. The male is velvety black, with a brilliant red head and hindneck, yellow thighs and underwing coverts, bright yellow eyes, and yellowish bill. The female is dull olive-green.

For his courtship displays, the male Red-capped Manakin chooses a straight, slender, more or less horizontal branch, which for a length of several feet is free of foliage and lateral branchlets, and is unobstructed by surrounding vegetation. Occasionally, a slender vine, stretched horizontally across a clear space, serves as his stage. Usually the manakins perform on branches twenty to fifty feet (6 to 15 meters) above the ground, but sometimes they are higher in the canopies of great forest trees. The courtship assemblies of Red-capped Manakins that I have seen consisted of no more than four or five adult males, stationed 20 to

125 feet (6 to 38 meters) apart, within hearing, if not within sight, of their neighbors. On these stations the manakins are to be found day after day throughout a long season.

As though his flaming scarlet head did not suffice to make him conspicuous amid the forest verdure, the manakin draws attention to himself by a surprising variety of vocal and mechanical sounds and odd antics. Most frequently heard is a sequence of notes that sounds like *psit psit psit p'tsweeee — psip,* the last note (which is often omitted) sharp and emphatic, whereas the whistled *p'tsweeee* is prolonged, high-pitched, and thin. The birds repeat these notes while they rest on their display perches during the hours of the day when they are least active. An exceedingly short, high *psit,* uttered singly or about five times very rapidly, is often heard from them. Very different is the high, shrill, rather harsh *tseeee* or *eeee* that a manakin emits as he returns to his perch after a short, circling flight, with this sound stirring all his resting neighbors to renewed vocal and muscular activity.

The male Red-capped Manakin, like the male *Manacus,* has enlarged, curved, and stiffened secondary feathers that appear to be responsible for the sharp sounds that he makes with his wings. As he darts back and forth between his display branch and a neighboring bough, he makes a single loud, sharp *snap,* like breaking a dry twig, each time he leaves his perch. While remaining on it, he produces a snapping *whirr* by beating his wings very rapidly. Or he may raise his wings to beat out a series of louder *snaps* more slowly. In addition to these explosive sounds, the manakin makes various whirring and rustling noises with his wings, either in flight or while perching.

The Red-capped Manakin's head is at all times so eye-taking that it is hard to imagine anything that he might do to make it more conspicuous. His display movements are, accordingly, largely such as expose his usually not-so-obvious lemon-colored thighs. Frequently, he stands on his display perch with his legs stretched up, thighs prominent, and about-faces as rapidly as he can. Keeping one foot in the same spot, he moves the other from side to side of his stationary foot as he pivots through 180 degrees, giving a resonant *flap* with each turn. At other times, he stretches up his slender legs, inclines his body so far forward that it almost touches the branch, raises his tail, often wags it from side to side, and, with steps so short and rapid that he appears to slide, advances tail foremost along his perch. After proceeding a few inches, he may turn around and glide in the opposite direction, always with his yellow pantaloons well exposed. Shaken by his innumerable mincing steps, the foliage at the twig's end vibrates rapidly. Or he darts swiftly back and forth

between his display perch and another a few feet distant. More spectacular is the circling flight, in which he flies out several yards from his
perch, loops around, and returns to it. As he nears the bough he stalls
momentarily while he makes a startlingly loud noise that has been compared to the sound of jerking taut a strip of strong cloth. With the shrill
eeee, he alights upon his perch.

Approaching silently, an olive-green female settles near the male's display perch, inciting him to perform with greater intensity some or all of
his customary antics, as though inviting her to come to the main branch.
If she accedes to his invitation, he courts her with feverish zeal. As he
glides backward toward her, displaying his yellow thighs, she may respond by moving toward him, tail foremost, rapidly beating her wings,
or she may sidle away from him. After more demonstrations, including the noisy circling flight, he launches once more into the air, veers
around, and hurtles toward the female with the loud *flap* and shrill *eeee*,
to alight upon her back for very brief sexual union. If he attempts to
repeat this, she may frustrate him by sidestepping just as he is about to
alight upon her. Soon she flies away, leaving him performing with energy
unabated after his strenuous courtship. During these nuptial trans-
actions, the chosen male's neighbors display vigorously in their own
places, or perhaps on a lower branch of his tree. They were never seen
to interfere.

The male most successful in attracting females spent most of his time
alone on his perch. His four neighbors in this assembly often met, two
by two, at points between their respective display perches, and similar
visits were witnessed at another assembly. On hearing an invitation
from one male, his neighbor would fly toward him, while he advanced
toward the other. Perching a few inches apart, the two males entertained
one another with subdued displays. The act most frequently practiced
on these occasions was an abbreviated version of the backward slide
with thighs exposed. These displays were not one-sided; after passively
watching his partner slide toward him, the recipient of this attention
often returned the compliment. These mutual demonstrations at points
between main display stations, which we also noticed among Orange-
collared Manakins, are not in the same category as the coordinated activities on main display perches to be described below.

The Golden-headed Manakin (*Pipra erythrocephala*) of northern
South America resembles the Red-cap, with the difference that its head
and hindneck are golden yellow, its thighs scarlet and white. The calls
and displays of the birds watched by David Snow in Trinidad differed
little from those of the Red-cap. The male practices the same backward

glide with hindparts elevated to reveal his colorful thighs, and he has a similar display flight, with the difference that he perches briefly between his outward flight and his return to his main perch in an S-shaped course. He does not snap loudly in the manner of the Red-cap.

During intervals of inactivity at an assembly of Golden-heads, two neighboring males spend much time sitting a few inches apart on a branch between their display perches, much as Red-caps do. Snow noticed that by consistently facing away, or half away, from each other, they reveal an ambivalent relationship, as though they craved companionship yet hesitated to become too intimate. When one slides backward toward the other, the latter moves an equal distance away, or perhaps flies to a nearby perch. Nevertheless, these manakins who sit close together spend more time in the assembly than others who remain alone. Although a manakin may, at separate times, sit with different partners, more than two never rest together simultaneously. Two who habitually rest together often fly off together, probably forage together, then return together.

Even smaller than the Red-capped Manakin, the tiny male Blue-crowned Manakin (*Pipra coronata*) is all velvety black, with a large oval patch of bright cobalt blue on top of his head. The female is fairly bright green. In forests from Costa Rica to Bolivia and western Brazil, they live chiefly in the understory, amid shrubs and small trees. In such situations males assemble for courtship, each in a poorly defined area about twenty or thirty feet in diameter, seventy-five feet or more from his nearest neighbors. The headquarters of each is a young tree with slender, horizontal branches, leafless except at the ends, where he spends much of his time from about six to thirty feet above the ground, delivering various combinations of a soft little trill and a harsh *k'wek*. The trill is frequently heard from both sexes; the *k'wek* rarely except from males in the assembly.

From time to time the Blue-crowned male flits back and forth between neighboring perches in his tree, or he makes looping flights between slightly more distant branches. While perching he flaps his wings, without making the sharp *snaps* of Red-caps. At intervals, he drops down to the undergrowth, to fly back and forth a few or many times, tracing an irregular zigzag course between slender upright stems, always within a yard or so of the ground. One manakin made fifty of these low flights with hardly a pause, then, after a rest, about as many again, always in the same small area, his territory.

Such wild darting to and fro often ends with an approach to the nuptial perch, a thin, more or less horizontal branch, often a fallen dead stick upheld by standing vegetation, about a foot above the ground in the darkest part of the forest. Dropping from a low shrub, he slants upward, to descend to this perch from a few inches above it, tracing a vertical sigmoid course. The moment he alights, he depresses his foreparts, bends down his head, and emits a little, harsh, grating sound while rapidly beating his wings. When a green female arrives, she flits back and forth amid the lowest shrubs, while the male darts about more obviously, much as when he is alone. Finally, she goes to his nuptial perch and remains while he flies up with his customary flourish and, fluttering his wings and voicing the usual grating sound, alights upon her back.

A peculiarity of the Blue-crowned Manakin's courtship is the separation of the nuptial perch from the tree where chiefly he calls and displays. This perch is so obscure that a female would hardly find it unless she had already spent some time attentively watching the male, who does not lead her to it but follows her there. The advantage of such a hidden perch may be that it screens the pair from other males who might interfere, and from predators.

The relatively simple displays of Blue-crowned Manakins include features that have been elaborated in other species. Among the latter is a social activity that develops when one male makes a friendly visit to another's territory. The two descend into the undergrowth and fly back and forth within a yard of the ground, much as the resident male more frequently does alone. Their paths often cross, with no indication of hostility. Sometimes three males indulge in these erratic flights together, and once, briefly I watched four so engaged. These social flights near the ground are different from the pursuits that ensue when one manakin invades another's display tree, starting a spirited pursuit in a circuitous course that may rise from the trees and shrubs of the underwood to the lower limbs of tall trees, all to the accompaniment of a slightly harsh *p'rrr* that expresses annoyance or anger. As in other manakins, these chases were never seen to end in bodily clashes.

A related species, the White-fronted Manakin (*Pipra serena*) of Venezuela, Guyana, Surinam, French Guiana, and northern Brazil is more ornate than the Blue-crowned Manakin and has a more coordinated social display. The male is velvety black with a silvery white forehead, bright blue rump and upper tail coverts, and yellow abdomen; the female is green. On grouped, defended territories 100 to 130 feet in diameter in the forest understory, males call incessantly on perches from three to sixteen feet high scattered through their domains. A male watched in Surinam by Prum called *whree* 7,400 times in slightly less than ten hours, during which he was present on his territory 72 percent of the time.

At intervals, the White-fronted Manakin flies to and fro between the branches on which he calls, much in the manner of Blue-crowned Manakins. From time to time he drops lower, to one of his courts, each about a yard in diameter, surrounded by five to ten upright saplings. Here he displays by flying rapidly across the court, alighting on the saplings facing inward in a rigid horizontal posture, often flicking his wings open to expose his bright blue rump. Or, holding his body nearly vertical and rapidly beating his wings in a buzzy, hummingbird-like flight, he hovers in shallow arcs back and forth over the court, barely alighting on a sapling before proceeding to the next one. More often than alone, the male White-front performs in this way with a neighbor who visits his courts. Coordinating their movements, the two take off and alight simultaneously, crossing each other in the air, often replacing one another alternately on a pair of perches. This dance is reminiscent of that performed by a male *Manacus* with a visiting female, but it is accompanied by neither mechanical nor vocal sounds. Moreover, White-fronted Manakins do not clear the ground of their courts, of which they may have, and use for display, up to five in a single male's territory.

The Crimson-hooded, or Orange-headed, Manakin (*Pipra aureola*) of eastern Venezuela, Guyana, Surinam, French Guiana, and Amazonian Brazil is more colorful than the preceding species of its genus. The prevailing black of the adult male's plumage is greatly reduced, giving way to crimson on his head, upper mantle, breast, and the center of the abdomen. On his forehead and throat the crimson pales to orange. His thighs and eyes are pale yellow; the bases of his remiges are white. The female is largely olive, as in related species. Brief observations by Snow in Surinam revealed a coordinated display by two males, one of whom joined the other on his main perch. While one waited there, the other flew to a branch about twenty yards away, then returned in the usual S-shaped course, making a soft *poop* (again the sound of cloth jerked taut) at the lowest point of his trajectory. This sound was the signal for the stationary bird to drop to a lower perch, sometimes after raising his head and calling *eeeew*, sometimes after simply ducking his head. This beautifully coordinated display continued for about three minutes, but, because of obstructing foliage, the watcher was not certain whether the same individual always flew out and returned, or whether the two alternated roles, as seemed probable.

Reversal of roles in a "swoop-in" flight, similar to that suspected of the Crimson-hooded Manakin, was definitely established for the Band-tailed Manakin (*Pipra fasciicauda*) by Mark B. Robbins, who studied this beautiful bird in southwestern Peru. Widespread east of the Andes, from

Peru to northeastern Argentina, this manakin has a yellow forehead, cheeks, and throat, scarlet crown, nape, upper back, sides, and chest, and a yellow abdomen. Elsewhere he is black, with white on his wings and a white band across the middle of his tail. He has the widespread pale yellow eyes of the male *Pipra*. The female is olive-green, more yellowish below. The courtship assembly in lowland rain forest that Robbins watched contained seven clustered display areas, each from 59 to 148 feet (18 to 45 meters) in diameter, with a main perch ten to sixteen feet (3 to 5 meters) high, plus three more widely separated territories. Associated with each territory was an extremely sedentary alpha, or dominant, male, and usually also a subordinate, or beta, male. In addition to these, two transient males, subordinate to both of them, might display briefly with them. When a female visited a territory, only the alpha male actively courted her. If he disappeared, his subordinate might inherit his territory.

The male Band-tailed Manakin advertises his presence by a variety of calls, some of which appear not to differ greatly from those of the Red-capped Manakin, plus a *klok* and a *kloop* attributed to his wings. His displays include short hops from side to side on his perch; short flights from perch to perch, with a *klok* as he alights on each; a rigid stationary posture; and a slow "butterfly flight" between perches that displays a broad band of white on each black wing and the white bar across the black tail. When another bird of either sex visits his perch, the owner may display to it by standing rigidly with raised tail. Turning his rear toward his visitor, he leans forward with lifted tail and rapidly vibrating wings, presenting a conspicuous area of yellow bordered by the black and white of the tail and wings. The bird's posture is much the same as that of a Red-capped or a Golden-headed Manakin performing the backward slide, but the Band-tailed Manakin has no contrasting yellow or scarlet thighs to show, and only rarely does he advance backward toward the visitor.

The Band-tailed Manakin's swoop-in flight is similar to the most dashing displays of Red-capped, Golden-headed, and Crimson-hooded manakins, and the sounds that accompany it are much the same. From his main perch the manakin flies to another fifty to a hundred feet (15 to 30 meters) away and a few yards higher. Here he alights with a *klok*, promptly turns around, and shoots back to his main perch in the familiar S-shaped trajectory, producing a *kloop* at its lowest point. Although he sometimes gives this display alone, nearly always he performs with a partner, usually another male in adult plumage familiar with the owner and his display site. By calls and a selection of his antics, he entices the

other, often his subordinate companion, to his perch. Then the two perform the full swoop-in flight alternately, exchanging positions on the
main perch as each in turn flies to the second perch and back again. This
closely coordinated, spectacular play may continue for several minutes,
to the accompaniment of the sounds that we have just noticed.

When an olive-green female Band-tailed Manakin approached a male's
territory, he invited her to his main perch with all his devices. His subordinate joined him in alternating swoop-in flights, the more to impress
her. The beta male gave other displays, as though helping his superior to
win the silent visitor, who did not display. When she went to the main
perch, the alpha male flew off, to return with a swoop-in flight. Turning
around in mid-air to face the way from which he had come, he alighted
on her back for a few seconds. When mating was finished, the female
flew away. Then the beta male, who had stood aside during the act,
rejoined the alpha male, and both gave a few advertisement calls. In six
months of almost daily watching, Robbins witnessed only this single
mating. The relationship of the dominant and subordinate Band-tailed
Manakins is similar to that among Blue-backed Manakins, in spite of the
great differences in their displays.

Unique Display of the Wire-tailed Manakin

The Wire-tailed Manakin (*Pipra* [*Teleonema*] *filicauda*) of northern South
America has a black back, rump, wings, and tail. His crown, nape, and
upper back are scarlet. His forehead, sides of head and neck, and all
underparts are golden yellow. A broad white band crosses his wings. A
curious feature of this yellow-eyed manakin is the prolongation of the
shafts of his tail feathers as fine, wirelike filaments that curve downward
and inward. The outermost of these filaments are the longest, often as
much as two inches (5 centimeters), which is half the length of the bird's
head and body. Toward the tail's center, these thin projections are progressively shorter. The function of the filaments, unique among manakins and with no close parallel in other birds, could hardly be imagined
until, in the late 1970s, two leading students of Neotropical birds, Paul
Schwartz and David W. Snow, solved the mystery.

For his displays, a Wire-tailed Manakin chooses an area of variable
size, about 65 to 115 by 33 to 82 feet (20 to 35 by 10 to 25 meters), amid
fairly open understory of light woods, often near an opening or a water-
course. Although never contiguous with a neighbor's territory, it may be
near enough to hear him or too distant. Within his display area, the

Wire-tailed Manakin, *Pipra filicauda,* male (left) and female.
Northern South America.

manakin performs on many slender, nearly or quite horizontal branches
about five to eight feet (1.5 to 2.4 meters) above the ground. Usually he
prefers one of these branches for his main displays.

The Wire-tailed Manakin's sounds, and with one important exception
his displays, are similar to those of the closely related Band-tailed and
Crimson-hooded manakins. He performs side-to-side jumps, a stationary
display, a butterfly flight, and a swoop-in flight with a *klok,* a *kloop,* and
a sharp *eeeo.* With a male visitor whom he entices to his display perch
with a variety of antics and calls, he joins in a coordinated performance,
the two alternately repeating the swoop-in flight, sometimes over and
over, in a dazzling exhibition that may continue from one or two minutes to five or ten.

The single display of the Wire-tailed Manakin that has no counterpart
in the antics of related species, the twist, provides the answer to the long-
standing question: What is the function of the tail filaments? For the full
realization of this unique performance, the manakin needs a partner,
who may be another fully adult male, an immature male, or a female.
Starting from a stationary display, the active bird turns his head away
from his passive partner, further lowers his foreparts, raises his tail, and
with vibrating wings pivots or twists his body from side to side through
an arc of about 60 degrees. As he hitches jerkily backward toward his
partner, he increases the tempo of his twists and raises his tail higher.

The partner, if experienced, advances toward the twisting bird until the uplifted filaments brush his or her throat, rhythmically and rapidly from side to side. In a well-coordinated display session, two adult males alternately tickle each other's throats. A young male partner sometimes tries to bite the shuttling tail, but eventually he lifts his head high enough to feel the filaments stirring the feathers of his throat. Or rarely, ignoring convention, an inexperienced male turns away to twist simultaneously with an adult, the two brushing their tail filaments together. A manakin whose invited partner is slow to approach may twist alone.

When a female enters a male's territory, he invites her to his main perch with frenzied calls, butterfly flights, and other devices. If she can be induced to alight upon it, he assumes the tail-up-freeze posture, presenting to her view a patch of bright yellow, crowned by black, and bordered on each side by white framed in black. When he begins to twist, she sidles close to him, as though eager to feel his filaments caressing her throat two or three hundred times a minute. Finally, he takes off for a swoop-in flight, from which he alights beside her before he hops onto her back. She is won by a combination of sharp calls, bright colors, spectacular antics, and tactile stimulation. This last element of the Wire-tailed Manakin's courtship repertoire appears to be without parallel in other birds, in all of which, as far as known, vocal and visual displays, or at most gentle bodily contacts, suffice to prepare females for mating. The backward slide with raised hindquarters, which we first noticed in the Red-capped Manakin, was doubtless the evolutionary prelude to the twist, the elongated tail filaments a subsequent development.

The Social Dances of Chiroxiphia

The four species of *Chiroxiphia* exhibit the most elaborate social displays that have so far been reported of manakins or other birds. The males of all four are black and sky blue, with larger or smaller patches of scarlet on the top or back of the head. On the Long-tailed Manakin (*C. linearis*), Lance-tailed Manakin (*C. lanceolata*), and Blue-backed Manakin (*C. pareola*), the blue is confined to the mantle. On the Blue, or Swallow-tailed, Manakin (*C. caudata*), it covers most of the body, set off by black sides of the head, wings, and outer tail feathers. The Long-tailed Manakin (which reminded me of a bird of paradise) has slender, greatly elongated, black central tail feathers. The corresponding feathers of the Lance-tailed Manakin are much shorter and thinner, tapering to sharp points. The central rectrices of the Blue Manakin are blue, with black ends protruding about an inch beyond the others. Those of the Blue-backed Manakin are not elongated. Bright orange or flesh-colored legs and toes add to the elegance of these stout little birds. Juvenile males are olive-green like females and take three or four years to acquire adult plumage, the first indication of which is red on the crown. Long-tailed Manakins inhabit mostly arid country from southern Mexico to central Costa Rica. Lance-tailed Manakins range from southern Costa Rica to northern Colombia and Venezuela. Blue-backed Manakins are found from southern Venezuela, Guyana, Surinam, and French Guiana to Rio de Janeiro. The Blue Manakin lives in humid forests from southeastern Brazil and Paraguay to northeastern Argentina.

For the Lance-tailed Manakin we have only incidental observations. Careful studies reveal close similarities in the social arrangements and coordinated displays of the other three species. Those of the Long-tailed and Blue-backed Manakins are so alike that we may describe them together, then notice how the Blue differs from them. The display perches of the first two species are thin branches or vines, nearly horizontal or arched, and devoid of lateral branches or leaves for much of their lengths. A single cooperating group may perform alternately at three or four such perches, all situated within a yard or two of the ground, usually amid a thicket of twigs and vines which make it difficult for a person to approach stealthily enough to watch the displaying birds, who are so shy that they disperse at the slightest disturbance.

Strangely, although they prefer to perform amid dense concealing thickets, Blue-backed Manakins, and probably the related species, pluck leaves from immediately surrounding vegetation, carry them to the display perch, and drop them. To such defoliation, which we recorded of *Manacus* and shall notice again in certain birds of paradise, several purposes have been ascribed. It may admit more sunlight to illuminate the colors of the displaying birds; it may make it more difficult for predators, attracted by the birds' sounds, to approach unnoticed; or it may remove obstructions to the displays. In addition to pulling off leaves, Blue-backed Manakins peck at the bark of their display perches, which their feet tend to wear smooth as they perch and hop over them year after year, as long as they are available.

In northwestern Costa Rica, Mercedes S. Foster, working with banded Long-tailed Manakins, learned that adult males join in couples, which she compared to those of monogamous birds whose pair bonds endure from year to year. If they avoid mishaps, the two male partners remain together not only day after day throughout a breeding season but apparently throughout the year. With rare exceptions, they display only with

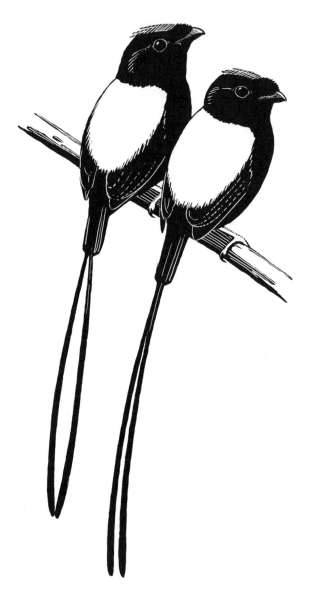

Long-tailed Manakin, *Chiroxiphia linearis,* males.
Southern Mexico to Costa Rica.

each other, usually at a single court; but their attachment to their partners is stronger than that to the court; occasionally they move together to another. Sometimes a third male is closely associated with the pair. Male Blue-backed Manakins associate in pairs or groups of up to eight or more mature and immature individuals. Two adults often sit closely side by side and utter their notes in unison, with no trace of the ambivalence that we noticed in males of *Pipra*, who rest farther apart and often turn their heads away from their companions.

To start a social display, a male Blue-backed Manakin, sitting alone in a tree, invites other males to join him, and perhaps females, too, with a rolling *chrrr*, often followed by a rather explosive *chup* that he may repeat once or twice. Or he voices the *chup* up to twenty times per minute. When, attracted by these calls, a second male joins the first, the two sit side by side and utter, sometimes for minutes on end, a series of perfectly synchronized ringing phrases, each consisting of a rapid series of one to five *chups*. The corresponding call of the Long-tailed Manakin sounds like *too-lay-do,* and is responsible for the bird's local name, "tolédo." Perched close together on their court (instead of above it as the Blue-back does), two Long-tails repeat this call in unison as frequently as twenty times per minute, often, with brief intermissions, for several hours together. It is rarely heard outside the breeding season, and almost never from single individuals. During the year's early months, this loud, ringing *tolédo* mingles with the clear *bob-white* of the Spotted-bellied Bob-white (*Colinus leucopogon*) and the melodious voices of the Banded Wren (*Thryothorus pleurostictus*) to impart a cheerful vernal atmosphere to the parched woodlands of northwestern Costa Rica.

After calling together, two male Long-tails or Blue-backs perch crosswise, a foot or two apart, on a low display perch, often surrounded by a few passive onlookers. One of the pair crouches a little, then flutters upward for one or two feet, legs dangling, bill and tail pointing downward. At the top of his ascent he hangs briefly, the red crown of his bowed head appearing unusually large and bright, his sky-blue mantle loosely fluffed. Meanwhile, his partner remains crouching, peering upward at him. The moment the first male alights at his starting point, the other rises in similar fashion. Thus, the two rise and fall alternately in rapid succession, each preserving his own place on the perch. A twanging or buzzing call, the guttural *miaow-raow* of the Long-tailed Manakin, the vibrant *aarr-r-r-r* of the Blue-backed, accompanies each jump, and guides one who recognizes it to a marvelous display in the secrecy of the thicket. As one watches the continuing performance, its tempo accelerates while the ascents become lower and lower, the pitch of the twanging calls rises higher and higher, until the jumps degenerate to seemingly uncontrolled flutters, the calls to a garbled buzz. The last jump of all is especially frenzied, the bird flapping from side to side as though helpless. Sudden sharp notes halt these wild flutterings; the actors immediately recover self-control and usually fly away.

When a female manakin, or in her stead a green immature male recognizable by a few red feathers on his crown, alights on a display branch, preferably at its upper end if it slopes, a different display ensues. Now the two adult male partners rest a little way apart, with their bodies parallel to the branch instead of transverse to it, always facing the green spectator. One is behind the other, and the one ahead, nearest the onlooker, flutters straight up into the air, uttering the usual twanging or buzzing call, and hangs there momentarily. Now the rear bird, crouching with eyes fixed on his companion in the air, creeps forward to the spot that the first has just left unoccupied. This bird now falls diagonally backward to the point left vacant by the second. Repeating these movements over and over, the two revolve in front of the green spectator, in what has been variously called a cartwheel or a Catherine wheel dance, all to the cadence of their unmelodious notes. Contrary to the impression made upon certain observers, Paul Slud perceived that female Long-tailed Manakins were intensely interested in this spectacle. Sometimes one became so excited that she hopped about between two performing males, spoiling the smoothness of their performance. As in the case of the straight up-and-down jumps, two or three sharp notes, quite different from those that accompany the dance, abruptly terminate the cartwheel display.

When the onlooker is a receptive female, one of the performers in the cartwheel dance moves to a neighboring perch to watch his partner court her with a quite different solo display, which David Snow has well described for the Blue-backed Manakin and which probably does not differ greatly in the Long-tailed. The courting male flutters around her with wings stretched outward, like a butterfly floating in the air, crossing and recrossing the display perch, frequently alighting momentarily. Resting near the female, he faces her, crouching with lowered head and vibrating wings to exhibit his blue mantle and his red cap with its long lateral feathers projecting sideward like two little horns. At intervals he flies to a special perch some twenty feet from the main perch, utters a low, twanging *quaaa,* then, with a soft but distinct click of his wings as he takes off, returns to the display perch to resume his bouncing flight. Except for these occasional sounds, the prenuptial display is silent. It culminates when the male alights beside the female, then jumps upon her back.

Foster learned that it is always the same dominant member of a pair (or trio) of males who mates with the female, while his partner stands aside. The subordinate may gain experience by participation in these exercises, and possibly inherit the display perches if he outlives his superior, who is probably always the older bird. Nevertheless, this seems a precarious reward for his prolonged cooperation, and raises questions, implicit in all social courtship displays, that we shall presently address.

The courtship of the Blue Manakin differs from that of the other two well-studied species of *Chiroxiphia* chiefly in the greater number of participants, which precludes the simultaneous calling and alternate jumping by only two performers and leaves the cartwheel dance as the principal display. In Paraguay, Foster studied a courtship assembly with six main courts scattered over an area of twenty-two acres (9 hectares). Each court contained a number of low, horizontal vines or unobstructed branches of small trees upon which the manakins displayed. In contrast to the situation in certain other manakins, all members of an assembly were free to use any of the courts. As long as no female was present, none was excluded from the display perches. At a single court, up to six males might be present together, calling and displaying. More often, a single sentinel male called steadily *ptuwa, ptuwa, ptuwa* . . . from a high perch in the center of the display area; or sometimes two males called alternately. These notes advertised the presence of a male at a court and attracted females. The appearance of a green female incited more rapid calling by the male who saw her first, and this drew more males who, with calls and whirring wing sounds, flew excitedly around the display area.

The cartwheel dance that ensues when a female alights on a display perch is closely similar to that of the related species, even to the sounds that accompany it, but it is typically performed by three adult males, although two may undertake it if another is not available, and four may participate in it, which seems unusual. Now, when the manakin nearest the female (number 1) flutters upward with a twanging *aarr-r-r*, number 2 creeps forward to occupy the point that number 1 has just vacated, and number 3 moves ahead to the place of number 2, leaving the last position in the row free for number 1 to drop backward into it. As soon as he alights, number 2 rises up to repeat the cycle, and so on. With their red caps glowing, the three manakins "form a whirling torch in front of the female, who perches motionless, watching them, or betrays slight nervousness by an occasional quick flick of her wings" (Snow). Just as in the more northerly species of Blue-backed Manakins, as the dance continues its tempo accelerates and the jumps become lower, the twanging less distinct. The revolving cartwheel seems to collapse, and, as the jumping birds move back to the end of the line, they pass low in front of their waiting partners instead of above them, as at the start of the dance.

The performance stops suddenly when one of the participants, instead of completing his upward jump, turns in mid-air to face the others and voices a sharp, penetrating *zeek-eek-eek*. At this command, the others crouch motionless, tails in the air, heads depressed to the level of the branch. After maintaining this frozen attitude for a few seconds, they fly to nearby perches, as the bird who gave the order has already done. The female stays on the display perch, and the dominant male begins to woo her with a solo display that differs only slightly from that of the Blue-backed Manakin, watched from nearby not only by his partners in the dance but often also by several other manakins who have been spectators of the whole drama.

Nearly a third of the thirty-seven solo prenuptial displays watched by Foster were not closely preceded by cartwheel displays. Either no partner was available, potential partners did not become involved, or, in two instances, males attempting to perform were chased away. The dominant male Blue Manakins, whose command halts the cartwheel display, perform the greater part of the matings, although they appear not to enjoy this privilege so exclusively as does the dominant member of a pair of male Long-tailed Manakins. Again, we are faced with the question: Why do manakins join as equals in performances for which the rewards are so unequally distributed?

Diverse Displays of Manakins

The foregoing accounts include all the manakins of which fairly detailed studies are available, but they cover only about a third of the species in the family. In these accounts I have given what appears, from my own observations or the descriptions of others, to be the typical or usual behavior of the birds in their courtship assemblies, omitting, to avoid prolixity, divergent behavior that has been recorded. To round out this chapter on the amazing displays of a most fascinating family of small birds, I shall, in what remains of it, describe very briefly the antics of a few species that are less well known.

In the forest along the Río Napo in eastern Peru, I met a very small manakin whose bright red cap, covering the whole top of his head, contrasted with the clear olive-green of his remaining upperparts. Below, he was largely chestnut, finely streaked with white. While I watched this Striped Manakin (*Machaeropterus regulus*), so different from any other member of the family that I had seen, he bent forward and, with widely open mouth, emitted a sharp, buzzing sound. Then he revolved rapidly all around a slender twig in the top of a medium-sized tree, while a much more plainly attired female or young male of his species looked on. Helmut Sick has described how a male clings, head downward, to a thin, vertical twig and turns so rapidly that he becomes a blur.

Not all male manakins are strikingly attired. Among the plain species is the flycatcher-like Wied's Tyrant-Manakin (*Neopelma aurifrons*) of eastern Brazil. Both sexes are olive-green above, grayish below, with yellow on the center of the belly. Their only adornment is the golden yellow or orange streak on their crowns, narrower on the female. The single display ascribed to the male is a weak call of three or four syllables, repeated as he sits alone on a low perch amid dense vegetation. The similarly plain Pale-bellied Tyrant-Manakin (*N. pallescens*) of Amazonian and eastern Brazil jumps upward and spreads his yellow crown patch while he makes a noise like *dop-dop* by striking together his wings, devoid of special modifications for producing sound.

To end this long chapter, let us glance at another very plain species, the Tiny Tyrant-Manakin (*Tyranneutes virescens*), widespread in northern South America. Only three inches (7.5 centimeters) long, it is olive above and grayish below, with a semiconcealed crown patch, much smaller on the female than on the male. David Snow found groups of two to four of these diminutive manakins scattered through a Guyanan forest. The calling perches of group members were about thirty to forty yards apart, near enough for the birds to hear the four-syllable *chuckle-de-dee* of their closest neighbors. This somewhat hoarse whistle was repeated with great regularity approximately every six seconds. Since one of these birds spent about 90 percent of the time calling, he must have repeated his little ditty nearly six thousand times in a day.

The Tiny Tyrant-Manakin's most prominent display was a floating flight with very rapid wing-beats, body half upright, legs dangling, neck stretched and thin, and crest fully erected like a spiky golden coxcomb. In this attitude the little bird made short sideward jumps along his perch and floated between neighboring branches, or sometimes for as much as ten yards. On these longer flights he made a sharp, mechanical trill not heard at other times. While perching between flights, he sometimes repeated rapidly a soft version of the *chuckle-de-dee*. With neck fully stretched and crest raised, he slowly swung his head from side to side while he continued to face forward. At the limit of each sideward movement he paused momentarily, as though craning his neck to see around an obstacle in front of himself, first trying one side and then the other.

Our prolonged, detailed studies of manakins are of social species that have colorful plumage, attract attention by loud sounds and curious displays, are fascinating to watch, and, on the whole, are more accessible to ornithologists than many species of remote forests whose habits may be no less interesting. On present information, it looks as though the manakins that have most claimed the attention of naturalists are just those that have been most favorable for the operation of sexual selection, which has promoted the evolution of their adornments and the postures and movements that exhibit them. This correlation may not be fortuitous; the courtship assemblies most interesting for us to watch may offer optimum conditions for female choice, the mainspring of sexual selection. But perhaps we should suspend judgment until we learn more about the neglected species, which little by little are becoming better known as naturalists have better opportunities to study them in South American woodlands.

References

Chapman 1935; T. A. W. Davis 1949a; 1949b; T. H. Davis 1982; Foster 1977; 1981; Gilliard 1959a; Prum 1985; 1986; Prum and Johnson 1987; Robbins 1983; Schwartz and Snow 1979; Sick 1959; 1967; 1984; Skutch 1949; 1967; 1969; Slud 1957; B. K. Snow and D. W. Snow 1985; D. W. Snow 1956; 1961; 1962a; 1962b; 1963a; 1963b; 1963c; 1971; 1976; 1977; Willis 1966.

9.
The Courtship of Cotingas

The cotingas are an amazing avian family of controversial limits. Some recent classifications remove such long-standing members as the tityras, monogamous birds who nest in holes in trees, the becards, also monogamous, who build bulky closed nests in trees, and the mourners, whose habits are little known. Even with these exclusions, the sixty-five species of cotingas, arboreal birds confined to continental tropical America, are an extraordinarily diverse family, including some of the smallest as well as largest of passerine birds. They are black, blue, green, yellow, red, and even wholly white, which is rare among the birds of the tropical forests. A few have colorful bare skin or wattles. Some are strangely silent; others are famous for the volume of their calls. Some breed as monogamous couples; others never join in pairs. Their nests include some of the slightest built by arboreal birds.

Among the controversial members of this family are the cocks-of-the-rock, who differ so greatly from all the others, in habits and anatomy, that the two species are sometimes segregated in a family of their own, the Rupicolidae. In some ways they remind one more of manakins than of cotingas, but with stout bodies and a length of eleven inches (28 centimeters) they are much bigger than any manakin and their coloration is quite different. Perhaps no other bird exhibits such large expanses of brilliant orange or orange-red as males of both species of cocks-of-the-rock. Both have large, upstanding crests of this color which, extending forward, nearly conceal their short bills. In the Andean Cock-of-the-Rock (*Rupicola peruviana*), which will first claim our attention, the tail and wings are largely black with unusually broad, pale gray inner secondaries that cover the rump. The females of both species are much darker, with greatly reduced crests.

The Andean Cock-of-the-Rock ranges through the Cordillera from northern Colombia to Bolivia, and from about four thousand to ten thousand feet (1,200 to 3,000 meters) above sea level. It eats mainly fruits, usually plucked in flight in the manner of many manakins and cotingas. Its habits were studied for fifteen months on the western slope of the Andes by César E. Benalcazar and Fabiola Silva de Benalcazar and reported in David W. Snow's monograph on the cotinga family. Here, in a wooded river gorge (the usual habitat of this species), six males displayed in pairs, the members of which had adjoining courts, each consisting of one or two branches or lianas, thirteen to twenty feet (4 to 6 meters) above the ground. Each court was about a yard wide, and the adjacent courts of pairs were separated from the display spaces of other pairs by twenty to thirty feet (6 to 9 meters). The six courts were within an area about eighty feet (25 meters) long. With bills and feet, the birds plucked nearly all the leaves from the horizontal limbs on which they performed.

The six males in the courtship assembly were ranked in a hierarchy. The bird, A, who was clearly dominant over all the others had the largest court and was the only one seen to mate with visiting females, who went almost exclusively to his court. He undertook to attack and expel all strange adult and juvenile males who approached the display area. Usually the first to arrive and the last to depart, he was present on his perches more than any of the other five. Moreover, he occasionally attacked his partner, grappling and falling with him if he resisted. These tiffs were apparently not serious, for, after a few seconds on the ground, the two would rise to their respective perches and continue to display close together.

The most frequent and conspicuous of the male cock-of-the-rock's activities was the "confrontation display" by members of a pair. The arrival of Male A on his court often started a bout of these displays. With a loud call, given while he bowed forward and often several times repeated, he announced his presence. With half-opened wings, he jumped upward several times, dropping back to the same spot, and he audibly snapped his bill. When his partner arrived, the confrontations began. At the boundary between their courts, pair members faced each other with their heads and spread tails depressed and their wings extended, each presenting to the other his bulging crest and dorsal surface as a curving expanse of deepest orange. While assuming this posture, the birds called *youii*. Performing simultaneously, the three pairs of males produced a volume of sound clearly audible hundreds of yards away. Even in the absence of a female, they might continue to display and call the whole time they were present in the assembly, then fly away to forage for a few minutes before returning. As in manakins, members of a pair sought their food together.

Singly, or in groups of up to five, females flew directly to Male A's court and flitted among its branches, briefly fluttering their wings between flights, making little vertical jumps, and uttering short, subdued calls. With abrupt movements, Male A would drive some of them away. If one resisted, he chased her from branch to branch within his court, repeatedly directing the confrontation display to her. Then, facing her about a yard away, he repeated brief, sharp calls. Silently, she inclined her body forward. Flying directly to her back, he half spread his wings over her, while his partner, still displaying, looked on in silence.

After insemination, the female Andean Cock-of-the-Rock goes alone to lay her two buffy, spotted eggs in the concave top of a nest shaped like an inverted cone, which with mud strengthened by fern rootlets she has attached to a shady riverside cliff. Often her nest is in a small colony. With great constancy, she incubates for twenty-eight days. Nourished by their mother alone with fruits, frogs, and lizards, the young remain in the nest for forty-two to forty-four days — like the incubation period, exceptionally long for a passerine bird. Between breeding seasons, the female continues to sleep in her nest.

The Guianan Cock-of-the-Rock (*Rupicola rupicola*) occurs in mountainous regions from Guyana, Surinam, French Guiana, southern Venezuela, and eastern Colombia to Brazil north of the Amazon, mainly at altitudes below five thousand feet (1,500 meters). The male's crest is not puffy like that of his Andean cousin but a flattened helmet that extends from his nape to beyond the tip of his bill with a semicircular outline, emphasized by a narrow stripe of deep purple slightly inward from the margin. His tail, orange like his body, is broadly barred with black. His broad inner secondaries are edged on the outside with long, silky filaments that wave in the slightest breeze. Each wing, overlaid by the silky fringe, is black, with a white speculum conspicuous only in flight. As though so much orange plumage were not enough of this color for one bird, his eye is bright orange-red, his bill and legs orange-yellow, and even his skin is orange.

Two display grounds in the Kanuku Mountains of southern Guyana were watched by C. Thomas Gilliard and later by David Snow. Situated on forested ridges, each was frequented by about five to ten adult males, who displayed on patches of bare ground from five to seven feet (1.5 to 2 meters) across, which might be almost in contact or more than thirty-three feet (10 meters) from their nearest neighbors. Unlike the courts of *Manacus*, these patches were not deliberately cleared by the birds; the litter covering them was simply blown away by the powerful wing-beats of cocks-of-the-rock landing on them or flying from them. The male owner of each court claimed perches in trees directly above it, where he passed most of his time.

Males arriving at an unattended display area utter a loud, buglelike *ka-waooh* or *ka-haaow,* which may serve to attract other members of the assembly. While interacting aggressively among the boughs above their courts, they squawk, caw, and gabble in a way that reminded Gilliard of barnyard fowls. They thrust their heads upward and forward with a sharp click that is apparently made by their bills, which are too well hidden by their crests to be seen clearly. When a female appears, they stop bickering confusedly in the trees and fly down to the ground, each to his own court, where he lands with a loud squawk, then briefly beats his wings rapidly above his back, exposing the white speculum, all of which makes him extremely conspicuous. Then, for minutes together, he crouches and freezes, with legs flexed, dorsal surface nearly horizontal, rump feathers and upper tail coverts flared out. He may hold his head normally or tilt it sideways, flat crest parallel to the ground and one eye looking upward. With gradual changes of posture, he seems to try to present his orange head and body as fully as he can to the females flitting about in the trees above him. The very different displays of the two species of *Rupicola* are closely related to the shapes of their crests.

While Snow watched from a blind soon after midday, the males in the assembly became more excited, and two of them dropped down to squat immobile on their courts. Alighting behind one of them, a female hopped up close, to lean forward and nibble the long, silky fringes of his wing feathers. Nervous in the presence of the blind, she moved away briefly, then returned to caress again the fringes of the motionless male — when the click of Snow's camera shutter sent her flying. While she was at the court, a curious whining squawking came continuously from the trees overhead, contrasting with the silence of the displaying males and creating an air of tense excitement.

For parts of six years, Pepper W. Trail watched courting cocks-of-the-rock in Surinam at four sites with, respectively, one, two, six, and an average of fifty-five court-holding males. He found these birds extremely wary, quick to fly up together at the least disturbance. These panics were caused by the approach of flying raptors, prowling carnivorous mammals, harmless animals, and, mostly, nothing evident to the observer. At the large lek, where courts were closely spaced, more than 90 percent of 832 mass flights, called "spooks," were apparently false alarms. When no danger was detected, the birds might return to their courts in as little as thirty seconds; when the flight was set off by an attempted capture, they remained absent for minutes, continuing to repeat their ringing *hey* alarm call. The only raptor known to attack successfully was the Ornate Hawk-Eagle (*Spizaetus ornatus*), who claimed two victims. The Collared Forest-Falcon (*Micrastur semitorquatus*) attacked more frequently but failed in twenty-eight attempts. Most surprising was the capture of two or three displaying male cocks-of-the-rock by a Boa Constrictor (*Boa constrictor*). Although many birds vociferously mob every snake, large or small, that they detect, cocks-of-the-rock either do not recognize these reptiles as enemies, or they become so absorbed in their courtship displays that they fail to notice the approach of the insidiously creeping enemy. Although the birds at the small leks were much less likely to fly up in alarm than were those at the large assembly, they were more frequently attacked by raptors, which supports the theory that one of the advantages of displaying in leks, rather than in solitude, is increased vigilance by many eyes.

Instead of streamside cliffs, the female Guianan Cock-of-the-Rock plasters her solid bracket of mud, strengthened by fragments of plants, to a vertical rock face, in a cave or a crevice in the fractured stony outcrops of the ancient, eroded Guianan Shield. Alone, she incubates her two eggs for twenty-seven or twenty-eight days, then attends her nestlings without their father's help.

Cotingas are mostly larger than manakins and tend to remain higher in trees, often in great remote forests, and to have sparser populations. For all these reasons, they have been less extensively studied. Much of what we know about their courtship and nesting we owe to the indefatigable patience of Barbara K. Snow, who has watched them in Trinidad and Guyana. In the Kanuku Mountains of the latter country, she studied the Calfbird, known also as the Capuchinbird (*Perissocephalus tricolor*), which is distributed through the forests of northeastern South America, in Brazil north of the Amazon, French Guiana, Surinam, Guyana, and Venezuela. The sexes of this fourteen-inch (36-centimeter) bird are similarly clad in chestnut-rufous feathers, with black tails and wings. The bare skin of their foreheads, forecrowns, and faces is blue-gray. Their dark bills and feet are large and strong. Two courtship gatherings at the foot of the mountains were in second-growth trees or understory trees beneath the forest canopy. The more thoroughly studied of these leks was attended by four adult males, each of whom claimed a bare horizontal branch thirty to thirty-six feet (9 to 11 meters) above the ground, and in view of each other from twenty inches to fifty feet (0.5 to 15 meters) apart. These four adults were present most of the daylight hours. Four

Calfbird, *Perissocephalus tricolor*. Sexes similar.
Northeastern South America.

others, whose calls and behavior suggested immaturity, visited the assembly at dawn and dusk and intermittently through the day.

The Calfbird is named for its voice. A solitary male utters only the first half of the *moo* call, which frequently summons the others to the assembly. When at least two are present, the full *grr-aaa-oooo* is heard, often from both simultaneously. Leaning slightly forward, the bird inhales air while uttering the *grr,* followed by the *aaa* while he stands upright, raises his tail, fluffs out his orange undertail coverts, and erects all the feathers of the anterior half of his body to form a hood around his bare face and forehead. With the final bellowing *oooo* he leans slightly backward, at the same time depressing his tail to display his curved orange undertail coverts as two bright globes against the tail's black upper side. Adult males usually stand back to back while they *moo;* or, if on the same perch, they turn away from each other to look in opposite directions.

To assert his claim to his display perch, the Calfbird assumes, by imperceptibly slow movements, a quite different posture. With body and neck stretched horizontally and wings dropped, he flattens all his plumage except his undertail coverts, which stand out conspicuously beside his uptilted tail. He holds this posture for a few minutes or nearly an hour, meanwhile keeping an eye upon his rival, often by twisting or tilting his head. He displays in this fashion on a branch that another male tries to appropriate too near his own. Here the contestants may remain motionless, two or three feet apart, for many minutes. In another aggressive posture, the male Calfbird puffs out his breast feathers, flattens his posterior plumage, and tucks his tail between his legs, making himself look like a pouter pigeon. If his hostile mood intensifies, he fluffs all his body feathers and becomes very rotund. While in the assembly, males frequently threaten or chase each other and occasionally fight. The persistence of their assemblies is proof that their tiffs are not very serious. Often they divert their aggressiveness to the twigs and leaves around their perches, pulling them off and carrying them from tree to tree before dropping them. Both parties to a dispute may pluck foliage.

Female Calfbirds appear not to utter the *moo* call nor to display like the males. They voice a rasping *waaaaaa,* sometimes followed by a short *aw,* somewhat similar to the half-*moo,* and a subdued *wark* as a contact call. Because the sexes look alike, it was not always possible to decide whether a silent visitor to the lek was a female or a male. On the two occasions when an apparent female alighted on the dominant male's branch, the three other males of the assembly came into or near his tree. She was sandwiched between him and another. Two females are sometimes closely associated, foraging together, visiting the assembly together, and building their slight, open nests of forked twigs close together; two nests were found only 230 feet (70 meters) from the displaying males. Each female lays a single khaki-colored egg, spotted and blotched with brown. Without an attendant male, she incubates it for twenty-six or twenty-seven days, and feeds the young in the nest for an interval at least as long.

The Screaming Piha (*Lipaugus vociferans*) is another cotinga that, despite its mode of courtship, has developed no difference in the appearance of the sexes. Nine inches (23 centimeters) long, both are gray, paler below, with dark bill, eyes, and legs. The loud, frequently repeated calls for which this bird is named ring through evergreen forests at lower elevations over much of South America east of the Andes, as far south as Bolivia and Mato Grosso in Brazil. Up to thirty calling birds may gather in an assembly spread over several acres; but groups of four to ten, the birds spaced about 130 to 200 feet (40 to 60 meters) apart, are more usual. Their somewhat ventriloquial *pi-pi-yo*, or *qui, qui, y-o*, audible through a thousand feet (300 meters) or more of forest, seems to lure a wanderer on and on through the woods, perhaps to spots where gold or rubber trees abound, thereby earning for them the name "goldbird" in Guyana and "seringuero" in Bolivia.

One Screaming Piha spent 77 percent of his day on his territory, calling from one or two to eight times per minute, rarely more. When calling he opened his mouth widely, revealing its orange interior. Neighboring males called alternately rather than simultaneously, which required careful timing, as the full *groo-groo, qui, qui, y-o* took approximately four seconds. Calling on rather thin horizontal branches from mid-height of the forest to the understory, a piha covered all parts of his territory. In spite of hours of watching, Mrs. Snow did not learn what happens when a female visits a calling male. Only a single nest of this widespread bird has been reported, and its contents were not seen. Like its relative, the Rufous Piha (*Lipaugus unirufus*), it probably lays a single egg in a thin nest barely large enough to hold it, in the fork of a slender branch well below the forest's canopy, where wind is not likely to toss it out.

Still another cotinga without the sexual differences in appearance that so often evolve in birds with leks is the more brilliant Red-ruffed Fruitcrow (*Pyroderus scutatus*), whose various races are widespread in South America but absent from Amazonia. This large (15 inches, 38 centimeters) cotinga is mostly glossy black, with red-tipped orange feathers covering its throat, chest, and the sides of the neck. In the northwestern race the lower breast and belly are almost solidly rufous brown, but in the southeastern race these regions are black with spots of this color.

A courtship assembly of Red-ruffed Fruit-crows in Venezuela was studied by Paul Schwartz, who died leaving unpublished notes and sound recordings which David Snow summarized in his monograph on the cotingas. In this gathering, seven or eight males regularly performed on perches within twenty feet (6 meters) of the ground and only about ten feet (3 meters) apart. In the dim light of early dawn, they started to repeat the deep, hollow, booming sounds, like the bellowing of a bull, which have earned for these birds the vernacular name "pájaro torero." While emitting this far-carrying call, the birds often approached within about a yard of each other, bent forward, and bobbed up and down with their red ruffs hanging away from their chests like bibs. When a bird stood upright, its ruff stuck out clear of its body, apparently pushed forward by the inflation of the air pouch that gives resonance to the bellow. Although the fruit-crows often chased each other, on the whole they tolerated their established neighbors but tried to drive away intruding males. When the assembly was visited by silent fruit-crows who appeared to be females but could have been young males, the resident males would all turn toward the newcomer and boom together. In this cotinga, voice and behavior rather than appearance identify the sexes. Mating was not observed; and little is known about the fruit-crow's breeding except that it builds an exceedingly slight nest of sticks in the fork of a high, slender branch, and vigorously defends it from hawks.

In the forests of the highest mountains of the Brazilian state of Rio de Janeiro and neighboring parts of São Paulo and Minas Gerais lives a cotinga with a very different voice. Eleven inches (28 centimeters) long, the male Black-and-Gold Cotinga (*Tijuca atra*) has no bright color except the contrasting yellow patch on each wing and his brilliant orange bill. The female is dull olive-green, with a yellowish belly, undertail coverts, and edges of her remiges. In a steep-sided valley in the Serra dos Orgãos, David Snow found a group of males calling day after day in the crowns of large trees rising above the forest canopy.

Although cotingas are more notable for the volume than the melody of their utterances, the Black-and-Gold is one of the exceptions. His pure, sweet whistles, long-drawn and plaintive, have deeply stirred those who have had the good fortune to hear them in the wild forests where these birds dwell. Often a prolonged, continuous chorus of these lovely notes pours at dawn from the high treetops, but it appeared that one dominant individual contributed most of the melody. In any case, with only brief interludes, calling continued throughout the day, and intensified when a female arrived. In early morning, late afternoon, and on dull, misty days, the cotingas performed in exposed treetops, but when sunshine poured into the valley they preferred high but less exposed perches where they were difficult to see. Mating was not observed, and no description of the single reported nest is available. It is almost certain that the olive-green females rear their broods with no help from the black-and-gold males.

Not the least of the contrasts in this unpredictable family is that between the stentorian voices of certain species and the rarity or absence of vocal sounds in others. To the loud-voiced species already mentioned might be added the large black umbrellabirds, distinguished by the males' thick, helmetlike crests that project forward over their bills and the long, extensible, feathered or bare wattles dangling below their throats. The three species, which replace each other from Costa Rica to Bolivia and the mouths of the Amazon, have deep, booming calls, not unmelodious at a distance, when they sound like the lowing of a bull. Adult male Amazonian Umbrellabirds (*Cephalopterus ornatus*) and Long-wattled Umbrellabirds (*C. penduliger*) proclaim their presence in widely spaced forest trees, but not so far separated that they cannot hear the loud calls of two or three others. This situation is often designated as an exploded, or dispersed, lek. To wide-ranging birds, it offers females a choice of nuptial partners, much as does a compact courtship assembly of birds with less powerful voices.

Renowned for their far-carrying calls are the four species of bellbirds that dwell in tall forests from Nicaragua to southern Brazil and northern Argentina. Most bell-like is the voice of the eleven-inch (28-centimeter) White Bellbird (*Procnias alba*) of Venezuela, Guyana, Surinam, French Guiana, and northern Amazonia. A long black wattle hanging from above the base of the male's black bill and his black feet contrast with his wholly white plumage. Perching at the top of a tree rising above the roof of a forest, he opens his mouth widely to emit a stentorian *kong kay,* said to be audible at a distance of a mile. More melodious is his long-drawn-out, silvery chime, *do-i-i-i-ing,* usually followed by a faint echo, that deeply stirs the spirits of wanderers who hear it ringing through the wild woodland from a hidden source.

The loudest notes of the other three species commonly lack the metallic timbre of the White Bellbird and sound more wooden. Notable among them is the Three-wattled Bellbird (*Procnias tricarunculata*) of southern Central America, from Nicaragua to western Panama. A male whom I watched for many hours spent most of his days on the tip of an ascending dead branch at the very top of a tree no less than a hundred feet tall, standing in a new clearing at the edge of a highland forest. How distinguished this twelve-inch (30-centimeter) bird looked when viewed against the blue sky, his pure white head, neck, and chest contrasting with the deep cinnamon-rufous of the rest of his plumage! Three long, wormlike black wattles, which seemed to get in his way, hung from the base of his bill, one above it and one on each side. Opening his mouth wide to reveal a cavernous black interior, conspicuous at a distance of fifty yards, he struck out notes that often sounded like *BUCK wheat,* the

Long-wattled Umbrellabird, *Cephalopterus penduliger,* male.

Three-wattled Bellbird, *Procnias tricarunculata,* male (upper left) and
female. Nicaragua to western Panama.

first syllable deep and loud, the second higher-pitched and softer. Without closing his bill, he often delivered a whole series of these notes in various combinations, such as *BUCK wheat BUCK wheat BUCK,* or *BUCK buck BUCK buck buck buck,* the *buck* a subdued version of his louder call.

As the bellbird called with his huge black mouth gaping widely, he leaned so far forward that he seemed about to lose his balance. As though to recover it, he often flew out horizontally for a foot or so, turned sharply in the air, and regained his perch, where he spread his brown tail and retracted his neck. From time to time, he sharply shook his head with its dangling wattles, apparently displaying them, although it looked as though he tried to rid himself of an annoyance. What part, if any, they played in courtship I never learned, for I saw little of the females, who are considerably smaller than the males, dull olive-green above, and on the underparts sulphur yellow, striped with olive-green. As in the following species, the sexes come together on visiting perches beneath the forest canopy.

The most thoroughly studied of the bellbirds is the Bearded (*Procnias averano*) of northern South America and Trinidad. Slightly smaller than the Three-wattled, the male has a brown head, white body and tail, black wings, and on his chin and throat a dense cluster of stringlike blackish wattles. The courtship assemblies watched for three years by Barbara Snow in Trinidad contained three or four adult males stationed on slight promontories at the heads of narrow, forested valleys. From high, exposed treetops where they enjoyed wide outlooks, the birds repeated their loud, explosive, far-carrying *BOCK,* sounding like a sharp hammer blow on a block of hard wood.

Much of the bellbirds' time, however, was spent on smooth, slightly drooping, uncluttered branches only fifteen or twenty feet up in small trees of the understory. Although they performed on branches of several neighboring trees, one of the lowest was preferred by the dominant male and the females who came to mate with him. At the approach of a visitor of either sex, he jumped from one branch to another, remained for a few seconds motionless with crouched body and fanned tail, then turned rapidly to leap in the opposite direction. When the visitor was an olive-green, striped-breasted female who could be enticed to his low perches, he displayed to her by crouching, lifting a wing, and extending the leg on the side toward her to show a patch of bare skin colored like his brown head, set amid the white feathers of his thigh. He called attention to this odd decoration by briskly preening the plumage beneath his uplifted wing about thirty times per minute. After each preening movement, he raised his head to look at her, and as he lowered it to preen again he showed her his brown crown. When finally she settled on the mating perch, he preened less frequently and turned his whole body along the branch to face her. Then, suddenly, he leapt with a loud, explosive *BOCK* and alighted on her back. The territory of the alpha male who won the females was coveted by associates ever ready to appropriate it. This led to frequent ritualized encounters between the male bellbirds, but they were never seen to touch each other.

In contrast to the loud-voiced male bellbirds, females are silent. With the exception of the Bearded Bellbird in Trinidad, little or nothing is known about their nesting. On that island, ten nests were found in cacao and other trees in cleared land close to forest. Each was a slight, open structure made of forked twigs which the female broke from living trees, and each contained a single egg or nestling. The light tan egg, mottled with brown, was incubated for twenty-three days by the female alone. The hatchling is covered with pale grayish white down unusually dense for a passerine bird. Although most frugivorous birds enrich their nestlings' diet with insects, young Bearded Bellbirds receive only fruits, regurgitated by the mother, during the thirty-three days that they remain in the nest.

To emphasize the contrasts in the mating habits of a family that tends to run to extremes, we turn from its most vociferous to its most silent members. Notable among them are the seven species of blue cotingas, widespread in tropical America, which give the family its name. Males of these middle-sized cotingas are turquoise to deep blue, with patches of rich purple variously distributed on their underparts. The very different speckled females are largely brown and gray. Singly, or in small parties of both sexes, blue cotingas often rest on the topmost branches of tall trees, where, in bright sunshine, the males are so beautiful that, no matter how many times one has seen them, one cannot resist raising one's binocular to gaze once more with delight tinged with wonder, for they are the most enigmatic of birds. Why do they wander so far over the forests, appearing and disappearing unpredictably in any locality? Why are they so voiceless? How do these lovely, silent birds court the females?

Although I have long been familiar with Turquoise and Lovely cotingas (*Cotinga ridgwayi* and *C. amabilis*) and have seen something of several other species, the only sounds that I ever heard from males of any of them were low, clear twitters, trills, or tinkles, sometimes approaching dry rattles, which when softest could have been vocal notes but were more probably made by the birds' attenuated outer primary feathers.

The sounds are heard almost continuously while the males fly on longer or shorter courses but not while they perch. From a female of this genus, I have heard only the agonized shrieks of a Lovely Cotinga whose nestling was attacked by an Emerald Toucanet (*Aulacorhynchus prasinus*). Unlike some of the larger members of the family, at least some of the blue cotingas lay two eggs in their slight open nests, which, of course, are attended by the females alone.

Equally reluctant to use his voice, if he has one, is the male Yellow-billed, or Antonia's, Cotinga (*Carpodectes antoniae*), whose plumage is everywhere white or palest gray. For four months I dwelt in view of the display trees of a solitary male, and frequently watched him closely, without ever hearing a note from him. He had three preferred trees, several hundred yards apart, two with dead branches rising above the canopy of the forest, the third in a clearing beside the forest. Much of the time he rested in his high treetops in silent inactivity, to fly at intervals in a deep catenary loop from one leafless branch to another in the same treetop. This, and a short sidling movement along his perch, were the only displays that I noticed. His white form was nearly always so conspicuous against the blue sky that vocal advertisement might have been a superfluous expenditure of energy. For none of the three species of white cotingas, which replace each other geographically from Honduras to western Ecuador, are further details of courtship or the nests known.

What, aside from the preferences of the females, which often appear to be capricious, could have led to the evolution of such bizarre masculine characters as the long, feathered pendants of umbrellabirds, the wormlike wattles of bellbirds, the bright blue foliaceous excrescences on the heads of Bare-necked Fruit-crows (*Gymnoderus foetidus*), and, perhaps most unexpected of all, the little bare patch of brownish skin on the thigh of the Bearded Bellbird; what could have pushed the cotinga family to such opposite extremes as the powerful voices of the bellbirds and the silence of blue cotingas and white cotingas? Apparently, when the vagaries of mutation produce in the male a secondary sexual character attractive to females of the same species, any further development of this character, acting as a supernormal stimulus, attracts them more strongly, with the result that little by little, as mutations accumulate, the character becomes so exaggerated, so useless if not positively impedimental, in the individual's struggle for survival, that ordinary natural selection would suppress rather than promote it.

The cotingas reveal, even more clearly than the manakins, the contrary effects of sexual and natural selection, especially in humid tropical forests, where predation on nests is severe but adults enjoy, for small birds, fairly long lives. With food so abundant in the breeding season that these largely or wholly frugivorous birds can satisfy their appetites in a few minutes and spend 80 to 90 percent of their active day in their courtship assemblies or on their nests, with predation on adults so reduced that males can wear colors that contrast strongly with the forest verdure and spend much time in exposed situations, they can afford to dress extravagantly and lavish their energy in displays. It is far otherwise with the females of these prodigal males. They must be unadorned and silent to avoid drawing attention to nests which they often make barely large enough to hold a single egg or nestling, or at most two. The young themselves tend to be undemonstrative and silent, receiving their meals without the eager clamor that we frequently observe at nests of passerine birds exposed to less heavy predation. Thus, we find sexual and natural selection operating in vastly different degrees upon the sexes of a single species, and pushing them to opposite extremes, the former making the males ornate, conspicuous, and often seemingly careless of predation, the latter making the females and their nests as plain and inconspicuous as they can possibly become.

References

Sick 1954; Skutch 1969; 1970; B. K. Snow 1961; 1970; 1972; 1977; D. W. Snow 1976; 1982; Trail 1987.

Color plate 3 (on preceding page).

Green Peacock, *Pavo cristatus*, male displaying. India and Sri Lanka.
Inset: Tremminck's Tragopan.

10.
The Courtship of Birds of Paradise

No other avian family has such a large proportion of lavishly ornamented species, nor such extraordinary adornments, as the birds of paradise. However, they do not owe this name wholly to the splendor of their
plumage. The first specimens to reach Europe were prepared by native
hunters and traders who apparently thought that the quite ordinary feet
of these perching birds detracted from the magnificence of their plumage
and, accordingly, removed them. Since, lacking feet, they could not
alight, they were believed, by those who had never seen them alive, to
pass their lives flying high above Earth in an aerial paradise. Here, so
the myth ran, the females laid their eggs on the backs of the males, to be
hatched by solar heat. This fairy tale is perpetuated in the name, *Paradisaea apoda* (the footless bird of paradise), that Linnaeus, probably not
without a smile, gave to one of the most ornate species.

Most of the forty-three species in this family are found in New Guinea
and neighboring small islands, whence a few extend to the Moluccas and
northeastern Australia. Most live high in the trees of humid forests, from
coastal mangroves to lofty mountains. In size they range from six to
forty-four inches (15 to 110 centimeters), including some very long tails.
In a few rather plain species the sexes are alike, but the species with
more elegant males have much duller, cryptically colored females. Although our knowledge of the breeding habits of these birds of remote
forests is still sadly deficient, it is known that some of the less ornate
species, including Macgregor's Bird of Paradise (*Macgregoria pulchra*)
and at least some of the manucodes (*Manucodia* spp.) breed in monogamous pairs, of which the female builds the nest and incubates the eggs
alone but is assisted in feeding the young by her mate.

Although observations are few, it is doubtful whether any of the more extravagantly adorned males takes an interest in nests, which are attended by the modestly attired females alone. Most of the known nests are bulky open cups, with a foundation of coarse sticks or moss and lined with leaves, vegetable fibers, or rootlets and tendrils. Several species build domed nests with side entrances, and the King Bird of Paradise (*Cicinnurus regius*) prefers a cavity in a tree. Birds of paradise lay one or two, rarely three, eggs, with a pale, often pinkish or yellowish ground color variously spotted and blotched. They hatch in about seventeen to twenty-one days, and the young remain in the nest for seventeen to thirty days. Although a few birds of paradise are mainly insectivorous, most species vary a largely frugivorous diet with buds, flowers, leaves, insects, and an occasional small vertebrate.

For geographical reasons, the courtship of birds of paradise has received less study than that of the New World manakins. One of the better-known species reminds us of manakins because, like a few of them, it clears a court on the woodland floor for its displays. One of the smaller members of the family, the Magnificent Bird of Paradise (*Diphyllodes magnificus*), is widespread at lower middle altitudes in the mountains of New Guinea and extends to neighboring small islands. The male is succinctly described by Thomas Gilliard as "a starling-sized, yellow-collared, orange-winged green black bird with long loosely coiled central tail plumes." The detailed descriptions of this and many other of the more ornate birds of paradise are necessarily so long that their perusal leaves but a confused vision of splendor. The female is olive and brown with barred underparts.

The courts and displays of Magnificent Birds of Paradise were described by Austin L. Rand and Gilliard. On a forested slope, a male removes all leaves, twigs, and small plants from a roughly circular patch of ground about fifteen or twenty feet (4.5 or 6 meters) across and deposits them in a windrow at the lower edge of the carefully cleared court. Only immobile logs, large branches, and exposed roots break the smoothness of the ground. In this bare area stand up to twenty slender young trees that are dying or dead because the bird has stripped the foliage from them, up to a height of twenty-five feet (7.6 meters). He has also plucked bark from their stems, leaving them frayed. To these activities he devotes much time, picking up fallen leaves and twigs and tossing them aside with a flick of his head, moving them again if they fail to reach beyond the court. The removal of so much foliage from above and around the court permits more light to illuminate his splendors and makes it more difficult for enemies to approach unseen.

While alone, the court's owner perches for long intervals near its edge, preening his feathers and repeating loud calls that might be audible to males at courts half a mile away. Frequently he displays by spreading and contracting his metallic sea-green breast shield while perching close above his court or clinging in a horizontal posture to the side of a sapling standing in it. One afternoon, when from a blind Rand watched a male clearing his court, a female arrived and perched nearby. Immediately, the male flew up to cling to a vertical sapling about a foot above the ground, whereupon she alighted on the same stem about a yard above him. He displayed by pulsing his breast shield, turned toward her. As she moved from sapling to sapling, he followed, always alighting low so that she could look down upon the shimmering pulsations of the plumage that covered his breast. Much of the time he repeated low, enticing or questioning calls. After this play had continued for about ten minutes, the female hopped down a sapling toward the male, who displayed below her. He intensified his efforts to dazzle her, spreading sideward his glossy golden yellow cape margined with orange-brown. When she approached still nearer, he abandoned his pose to hop up and mate with her. After coition, he dismounted, raised his tail straight upward, and vigorously pecked her nape. After each peck, he drew back and opened his mouth widely to show her its yellowish green interior.

The reader may recall that hermit hummingbirds expose colorful mouth linings as part of their courtship displays, and we shall meet with other instances of this behavior among birds of paradise. Is this just an added touch to the male bird's exhibition of his regalia, the persistence of infantine behavior, or perhaps a suggestion that the female should soon be placing food in the similarly colorful mouths of her nestlings?

The related, equally embellished, but quite differently colored Wilson's Bird of Paradise (*Diphyllodes respublica*) lives in the hilly interior of islands off the western end of New Guinea. Its courtship has apparently never been studied in the wild, but in captivity the male clears a court on the ground, and when displaying expands his pectoral shield and opens his mouth to reveal its light green lining. Another genus that displays above bare patches of ground is *Parotia*, with four species distinguished by six long, naked feather shafts, each terminated by an expanded disk, that spring, three on a side, from the head, giving these splendid creatures the name six-wired birds of paradise. Although these birds are considerably larger than the Magnificent Bird of Paradise, their bare courts, as far as known, are smaller. A prominent feature of their displays is swaying the head to wave the "flags" at the tips of long, thin stalks.

In montane rain forests of Papua New Guinea, the Pruett-Joneses found Lawes' Six-wired Bird of Paradise (*P. lawesii*) displaying at courts that were mostly grouped in leks, although about a third of the displaying males were solitary. Each court was an area of the forest floor from which its single occupant carried all leaves, bark, and other movable litter each time that he visited it, keeping it clean and bare. He also plucked leaves from saplings on or around it. A single male might have from one to five of these bare patches, from which he expelled other males of his kind. About twenty inches (50 centimeters) above each court was a thin, more or less horizontal branch or vine, on which the owner displayed and mated. A unique feature in the behavior of this bird, which distinguishes it from other birds of paradise and from manakins that display at bare terrestrial courts, is the collection of objects, including shed snakeskin, scats of small mammals, bits of chalk, fur, feathers, and fragments of bone, mostly the first three. Arriving with one of these objects, the male bird holds it in the end of his bill while he rubs it methodically over his display perch, sometimes continuing for twenty minutes if he has found several pieces of snakeskin. This makes the perch smooth and shiny, and, when he uses fragments of chalk from exposed calcareous outcrops, leaves wide white streaks visible at a distance. After rubbing, he drops the object to the bare ground.

Male Lawes' Six-wired Birds of Paradise do not hold these objects while displaying to females. Throughout the day, they rearrange them on the ground, of their principal or most-used court if they have more than one. Usually they gather them toward the court's center, in clear view of the display perch. These fragments attract other individuals of the same species of both sexes, who are chased away by the owner when he is present but carry them off when he is absent. Pilfering males take the stolen objects to their own courts, rub them over their perches, and arrange them in the usual fashion. Females return repeatedly to appropriate all the snakeskin they can find, as well as other items. Probably they add the flexible skin to their nests, as birds of diverse families do, but this could not be confirmed because the few known nests were so high. Chalk fragments and mammal scats probably provided minerals for the females' eggshells. Apparently, collecting an abundance of materials for females to carry away does not increase a male's success in mating with them. He appears to bring these items primarily to polish his perch, behavior reminiscent of bower-painting by male birds holding wads of fibrous stuff in their bills. Polishing or painting, no less than the accumulation of objects on the courts, helps to bridge the gap between birds of paradise and bower birds. Incidentally, they add to the growing number of known avian tool-users.

Among the many birds in this family named for European royalty is the King of Saxony Bird of Paradise (*Pteridophora alberti*). The male is a thrush-sized black-and-orange bird, from the back of whose head spring two plumes twice as long as his body. Each consists of a long, wirelike shaft bordered on one side with up to forty-four little lobes or lappets that appear to be painted with sky-blue enamel. In New Guinea's central mountains, these fantastic birds live in cloud forests, chiefly from five thousand to nine thousand feet (1,560 to 2,750 meters) above sea level. Plainly clad females, who bear only a long, spikelike gray feather springing from the crown behind each eye, and young males are generally common within this altitudinal zone, but the males are hard to find unless one knows where they gather to display. The locations of these assemblies are known to the natives, who are reluctant to disclose them because the birds' enameled plumes are so valuable. Fortunately for Gilliard, he was able to persuade a Papuan helper to guide him to one of these leks in the midst of the forest. Here he found a number of interacting males, each of whom performed on an exposed, spirelike limb with a wide outlook, a thin, horizontal branch beneath a sparse canopy of foliage, or a high vine. These stations for display were from sixty to one hundred feet (18 to 30 meters) above the ground. Each bird was four hundred or more yards from his nearest neighbors. Although this appears to be a very wide separation for birds in a courtship assembly, these males were evidently within hearing of each other, for their calls, beginning with a prolonged hiss like the sound of escaping steam and ending with an explosive rasping note, carried for nearly a mile over the canopy of the cloud forest. Accordingly, these King of Saxony males, so difficult to find elsewhere, might be considered to form a true social gathering, an exploded lek, serving the same purposes as the more concentrated assemblies of Ruffs, hummingbirds, manakins, and other birds with less powerful voices.

Male King of Saxony Birds of Paradise pass much of the day on their display perches, uttering a thrice-repeated, gargling *grrrrreeaaa,* followed immediately by a low *ca-ca-ca* given twice. These notes with a hissing quality were emitted with the green mouth half open and pointing upward; it remained open between utterances. By swinging their heads forward and backward, the birds raised and lowered their long plumes, making them more conspicuous. A male that Gilliard watched performed on widely separated perches high in the forest. When a female arrived, he rested on a thin twig, which he caused to bounce up and down with him, while he repeatedly bowed forward and downward, all of which kept in constant agitation the head plumes that he raised

King of Saxony Bird of Paradise, *Pteriodophora alberti,* male.
Mountains of New Guinea.

steeply above his back. While he bounced, bowed, and loudly hissed, the
visitor approached so close that his blue enameled plumes swept in front
of her eyes. As the display promised to reach its climax, the male appeared to lose his grip on his swinging perch and fell with a harsh, gur-
gling hiss. Startled by this mishap, the female fled. With his head plumes
trailing behind his tail, he followed her closely until both were beyond
view. Gilliard knew nothing of the nest and eggs of the King of Saxony
Bird of Paradise.

Paradisaea is the genus for which the family is named and, with seven
species, the largest of its twenty genera. In size, these big birds of paradise range from about eleven to eighteen inches (28 to 46 centimeters),
not including the adult male's greatly elongated, wirelike or ribbonlike
central pair of tail feathers. A distinguishing feature of this genus is the

dense tuft of long, filmy plumes, white, red, orange, yellow, or blue, which springs from each side of the adult male's breast and is the outstanding glory of his displays. The smaller, plainer females lack these tufts and tail wires. Unlike many other members of the family, these birds live mainly at lower altitudes, from the coasts of New Guinea and neighboring islands up, in several species, to about five thousand feet (1,525 meters), although one, the Blue Bird of Paradise (*P. rudolphi*), inhabits the mountains of eastern New Guinea from about 4,500 to 6,300 feet (1,370 to 1,920 meters). This distribution has made *Paradisaea* more accessible to naturalists than are species confined to the rugged interior, long the stronghold of fiercely warlike tribes, and in consequence they are, on the whole, better known, although much remains to be learned about them. They are among the more social of the birds of paradise, and their often noisy courtship assemblies attract attention.

Largest member of the genus is the eighteen-inch (46-centimeter) Greater Bird of Paradise (*P. apoda*) of southern New Guinea and the Aru Islands between this great island and Australia. Largely maroon-brown, with an orange-yellow crown and hindneck and iridescent oil-green throat and foreneck, the male would be a handsome bird even without his massive tufts of lacy plumes, yellow with pale cinnamon ends and up to twenty-two inches (56 centimeters) long. His wirelike, dark brown central tail feathers, up to thirty inches (77 centimeters) in length, extend well beyond his plumes. The very different females and juvenile males are mostly vinaceous maroon. Both sexes have lemon-yellow eyes and dull brown feet.

On his visit to the Aru Islands in 1857, the great naturalist-explorer Alfred Russel Wallace was almost as eager to learn the habits of the Greater Bird of Paradise as to obtain the birds themselves. Unfortunately, painful ulcers on his feet, resulting from an excessive number of insect bites, prevented his walking, and he depended largely upon his hunters and local people for information about the birds. In his classic account, he told how they danced "in certain trees of the forest, which are not fruit trees as I first imagined, but which have an immense head of spreading branches and large but scattered leaves, giving a clear space for the birds to play and exhibit their plumes. On one of these trees a dozen or twenty full-plumaged male birds assemble together, raise up their wings, stretch out their necks, and elevate their exquisite plumes, keeping them in continual vibration. Between whiles they fly across from branch to branch in great excitement, so that the whole tree is filled with waving plumes in every variety of attitude and motion." He found these birds numerous; daily he heard their *wawk-wawk-wawk-wŏk, wŏk,*

Greater Bird of Paradise, *Paradisaea apoda,* male displaying.
Southern New Guinea and Aru Islands.

wŏk, so loud and shrill that it carried a great distance and was "the most prominent and characteristic animal sound in the Aru Islands."

Surprisingly, Trinidad and Tobago, the country where so much has been learned about the courtship of manakins and Bearded Bellbirds, has also contributed most of what we know of the details of courtship of this bird from the opposite side of Earth. Early in this century, when many thousands of birds of paradise were annually slaughtered to adorn the hats of European and American women, Sir William Ingram, founder and first editor of *The Illustrated London News,* aviculturist and conservationist, feared that they faced extinction. To save the Greater Bird of Paradise from such a fate, he bought Little Tobago Island, an uninhabited wooded islet a mile offshore from Tobago, and in 1909 stocked it with forty-four birds collected for him in the Aru Islands. In their new home, the birds of paradise survived and nested but failed to become numerous. In 1958, when Gilliard and the photographer Frederick Truslow watched them for three weeks, fifteen individuals were certainly present on the islet, and possibly as many as thirty-five. During a nine-month study in 1965–1966, James J. Dinsmore could find only seven birds of paradise, including four males in full adult plumage, one sub-adult, and two in the plumage worn by both females and young males. Possibly a hurricane two years earlier had reduced the population.

Dinsmore found four main display stations, each a tree or group of trees frequented by one or more males. Three of these stations were within 207 feet (63 meters) of each other on a wooded hillside, not too far apart for the single plumed male who regularly attended each of them to hear, and probably often see, his neighbors. Each of these males was often visited by the others, who tended to remain somewhat apart from the horizontal or gently sloping, leafless branch, beneath a fairly thick canopy, where the resident male habitually displayed. The fourth male, whose short side plumes suggested that he was younger than the others, had a much more distant display station where he was infrequently seen. Dinsmore never saw the birds pluck leaves from their display branches, but, with harsh nasal calls, they often tore pieces from the large, fan-shaped fronds of nearby palms.

Early in the season, the males on Little Tobago often displayed together, much as, a century earlier, Wallace had described for larger groups in the Aru Islands. Later in the season, and much more frequently, they displayed alone, each in his own space. One male regularly permitted others to alight on his display station but consistently chased them from his main display branch. This bird repeatedly drove the female from this branch, once five times in succession, until by her persistence she demonstrated her readiness to mate and was accepted. All of the six mountings that Dinsmore witnessed were by the male with the longest, most richly colored side plumes, who was the most active in display, at the most central station. Three of these completed or attempted matings occurred when no other male was present.

So much for the social organization of the Greater Bird of Paradise in a depleted population, the one that has been most thoroughly studied. Before looking at his lovely displays, which have much in common with those of related species, let us see how other members of the family arrange their courtship, beginning with Count Raggi's Bird of Paradise (*P. raggiana*), more recently called the Raggiana Bird of Paradise. With a length of thirteen or fourteen inches (33 or 35.5 centimeters), the male is considerably smaller than the Greater male. His body, wings, and tail are mostly brown to blackish. The yellow of most of his head extends as a complete collar around his neck, below his iridescent oil-green throat. His greater wing coverts are partly yellow. His thick tufts of lacy side plumes are blood red to apricot orange, becoming pale rose-cinnamon on the outer third. The female is much plainer. Confined to eastern New Guinea, these birds are abundant not only in upper levels of the forest and at its edges but also in isolated groves of trees amid grassland or native gardens. Although small bands of females and young males are frequent, adult males are difficult to find except in the vicinity of their courtship assemblies.

Gilliard found Count Raggi's Birds of Paradise displaying in assemblies of about three to six adult males in isolated groves of tall, slender trees. In a space of about one hundred to four hundred feet (30 to 122 meters), the display stations of individual males were from twenty to fifty or more feet (6 to 15 meters) apart, high up under the canopy, one to a tree, or sometimes several in the same widespreading crown. The birds performed chiefly in the early morning and sporadically through the rest of the day. They were noisy birds, whose loud calling while they displayed carried far. In addition to vocal sounds, they made resounding thuds by clapping their wings together above their backs, while they crouched low upon their perches with their long flank plumes expanded upward. They did not use their voices while they clapped.

Occasionally, a male became curiously agitated, perhaps because another had encroached upon his private display space. With his raised plumes nearly concealing his body and head and calling noisily, he charged back and forth on his display limb. Neighboring males left their own spaces to gather around him, amid a confusing crowd of younger males and females. After a native plume-hunter shot a displaying male, his space remained vacant for two mornings — evidence that it had been his private domain. On the third morning, a young male in female plumage jumped onto this display perch, crouched, and clapped his wings above his back just as adults did.

In later years, Bruce Beehler and his assistants spent more than 150 hours watching nine leks of the Raggiana Bird of Paradise situated in large expanses of continuous forest or in remnant patches near forest edge, all in Papua New Guinea. At each location from one to ten adult males displayed on upper limbs of a tall tree. The rare instances of aggression among males were usually mild scuffles or chases, very seldom fights in which the participants clutched and fell together from the tree. In the early morning, when the males displayed most actively, females arrived in pairs or small parties of up to five or six. Approaching a male of her choice, a female often alighted on his back, thereby signifying her readiness to mate with him. After displaying his gorgeous adornments, he mounted her. Matings were more equitably distributed among these male Raggiana Birds of Paradise than they are among some other birds with leks (see Table 2 in Chapter 14). At the assembly most thoroughly studied, the four males performed, respectively, fifteen, four, seven, and three matings. Very rarely one interfered with another who was displaying to a female close beside him.

In Papua New Guinea, several nests of Count Raggi's bird have been found, one in a botanical garden, another in a rubber plantation. One nest was a substantial cup of vines and dead leaves, lined with brown fibers, mostly from palm fronds. Each of two nests contained two eggs of a beautiful pinkish cream color, streaked with rufous and gray. Bruce Beehler and S. G. Pruett-Jones learned that 90 percent of the diet of Count Raggi's bird consisted of fruits, the remainder of arthropods.

The Lesser Bird of Paradise (*Paradisaea minor*) resembles the Greater but is smaller. In northern New Guinea this beautiful bird frequents tropical rain forests, swamps of sago palms, and patches of woods amid grasslands, as well as trees in native gardens and on the outskirts of coastal towns. Its noisy parties usually remain in the upper half of the forest. Gilliard watched parties of from five to fifteen males moving about in a confusing manner and displaying, in the top of a single tree or in half a dozen adjoining trees, all to the accompaniment of a medley of loud calls. They did not appear to have individual stations. Occasionally, one tore off leaves as though angry. Gilliard's observations were apparently made early in the season, before the birds had chosen display stations and settled down, for in a later year Bruce Beehler found eight males displaying on horizontal or gently sloping branches of a lofty tree beside a clearing, each on his own private perch, which might be as close as twenty inches (50 centimeters) to that of a neighbor. Even in the owner's temporary absence, one male did not intrude upon another's station. When a female arrived, she went from one male to another, as though appraising them, before she invited the preferred bird to mate with her. With one exception, all of the twenty-six matings that Beehler witnessed in forty-nine hours were performed by the centrally located male, who spent much more time on his display perch than any of the others. The male who was the next most constant attendant mated once. While the chosen male mated, the others did not interfere but remained on their own perches. Immature males in transitional plumage, who spent most of their time calling in trees scattered through the surround-

Lesser Bird of Paradise, *Paradisaea minor,* male displaying.
Northern New Guinea and western Papuan islands.

ing forest, gathered around to watch the proceedings whenever court-ship became most active, but they neither intruded nor displayed. Most matings occurred in the early morning or midafternoon.

The coloration of Goldie's Bird of Paradise (*P. decora*) does not differ greatly from that of the Greater and the Lesser, except that the male's long, lacy side plumes are mostly bright red, tipped with grayish buff. This species is found only on Normanby and Fergusson islands off the southeastern tip of New Guinea, mostly in the hill forests above one thousand feet (300 meters). Here, Mary LeCroy and her associates watched an assembly of eight or ten plumed males, who displayed in four tall, straight trees spaced in a rectangle of about one hundred by fifty yards (92 by 46 meters) in mature tropical forest. The birds per-formed sixty or seventy feet (18 or 21 meters) up, near the bases of branches in the mid-canopy, beneath rather sparsely foliaged crowns. The plumed males regularly plucked leaves from around and above their main display limbs, once seventeen of them in quick succession, drop-ping each in turn. Sometimes one bit off a twig with several leaves, or held an attached twig beneath a foot while he tore away the leaves one by one.

In addition to the mostly loud, unmelodious calls typical of birds of paradise, Goldie's birds duet. Standing face to face, usually from four to ten feet (1.5 to 3.5 meters) apart, with their side plumes raised above their backs, often in the presence of a female, two males utter alternately a loud, metallic *waak*. As the performance proceeds, it accelerates until the notes merge into a continuous metallic rattling, which rings afar through the forest and reveals the presence of an active assembly. While they duet, one or both performers may run up and down the branch.

Early one morning, while a pair of males duetted, two females ar-rived. One of them settled on the horizontal display limb between the duettists; the other perched nearby. Several plumeless young males who hopped restlessly around them were chased away a few times by the adult males but mostly ignored by them. After the displays and duetting of the two adults reached a climax, one of them moved aside, to sit qui-etly on a neighboring branch while the other continued to address the female on the display limb. His movements slowed until he was almost static; no more vocal sounds were heard. Several times the female left the main display perch but returned almost immediately. Finally, she began to solicit with quivering wings while standing near the adult male, who for five minutes continued his slow, rhythmic display. Taking ad-vantage of the adult male's slowness to respond, one of the displaying

young males who hovered close around the courting pair, no longer op-
posed by the adults, moved in and mated several times with the female.
Once two of them mounted her in succession. In each case, coition lasted
only a few seconds and was possibly incomplete, as the female never
ceased to solicit from the plumed male. Not until he had displayed to her
for about half an hour did he begin to hop stiffly up and down near her,
coming closer until he laid his neck and breast over her back and rubbed
it. Then he mounted, with his wings around her body while they mated.

After this episode, all the birds remained in the tree, and the whole
sequence was repeated, starting with duetting and joint displays by the
two plumed males. Again, the plumeless males were chased away during
the duetting and joint displays but were tolerated during the interval of
intense display by a single adult male. Again, these young birds mounted
the soliciting female briefly several times, before the plumed male
mounted and embraced her as before. The observers were not certain
whether the male who now mated with the female was the same adult
who had done so earlier.

These extraordinary observations, which I have retold with slight
abridgment, raise questions vital to the theory of sexual selection. Why
did the adult male permit the subadults to precede him in mounting the
female whom he was fervently courting? If, as it appeared, she had chosen the plumed adult to sire her nestlings, why did she accept the plumeless interlopers? A possible explanation is that the adult male had
worked himself into such an ecstatic state, and the female was so absorbed watching him, that neither paid attention to the young males.
Another possibility is that they were ignored because they would not
succeed in inseminating the female (although plumeless birds of paradise
have sired nestlings in the very different conditions of aviaries), and a
long racial experience of such fruitless intrusions had made the chief
actors careless of them. Or was this simply a very abnormal situation? If
it were frequent, the splendid plumes that birds of paradise took such
long ages to acquire would degenerate for lack of the selection that produced and maintains them. We shall return to this question in Chapter 14.

Noteworthy is the fact that while a male Goldie's Bird of Paradise
courted and mounted a female, the partner with whom he had been
displaying and duetting stood aside and quietly watched without interfering. Similarly restrained behavior by a duetting partner and two other
plumed males has been reported of Count Raggi's bird. This corresponds
closely to the conduct of manakins, especially *Chiroxiphia,* in similar
circumstances.

Somewhat higher in the mountains of eastern New Guinea lives the

Blue Bird of Paradise (*P. rudolphi*), regarded by some as the loveliest of all. The jay-sized black male has blue wings and long blue, purple, and cinnamon side plumes. His two very long and narrow black tail streamers often bear subterminal spots of blue. His blue eyes are set in interrupted white rings. The female of this species lacks ornamental plumes, but in coloration she resembles the male more closely than do the females of other birds of paradise. Instead of assembling in leks, male Blue birds display alone at stations well separated along forested ridges, where each attracts numerous females. The Pruett-Joneses found one nest high in a tree, where the solitary female incubated her single egg and attended her nestling without a male's help. After the loss of this nestling, she tore her nest apart, as, in tropical American forests, female cotingas frequently do after a failed or successful nesting.

The superlative beauty of birds of paradise is most fully revealed in their courtship displays, to which chiefly they owe their fame. Many species have been kept for years in private aviaries and public zoological gardens, where the males' demonstrations can be watched and photographed much more satisfactorily than in the crowns of tall tropical trees. Accordingly, a substantial part of what we know about these displays has been gathered in these artificial conditions, where the full sequence may not be shown.

As one would expect, in all seven species of *Paradisaea* the long side plumes are the central feature of the male's displays. While he rests or forages, they trail behind him; when he begins to perform, he elevates them above his back. However, he does not at once assume his most captivating pose but works himself up to the climax by a sequence of preliminary movements, which vary somewhat not only from species to species but also in different performances of the same species or individual. Nevertheless, they have many features in common. The visual displays are preceded and accompanied by much loud calling, exceptionally by synchronized duetting, as in Goldie's Bird of Paradise, which advertises the assembly's location to females scattered through the forest.

In his prolonged study of the Greater Bird of Paradise on Little Tobago Island, Dinsmore recognized five phases of the courtship display. Phase 1 is the "wing pose," which may be assumed in the presence of a male or a female, or in the absence of any other bird. Arriving on his display branch, the male extends his wings rigidly in front of his body, raises his side plumes above his back, and brings his tail forward beneath the perch, all while rapidly calling *wauk*. After holding this pose for several seconds, he gradually drops his wings to his sides and flaps them vigorously. With a female near but not upon his display perch, he may repeat this sequence every ten or twenty seconds for half an hour or more.

In phase 2, which Dinsmore called the "pump display," the male bird of paradise crouches with his body nearly parallel to his perch, his spread wings cupped slightly around it, his head and bill pointed downward, his plumes standing almost straight up. In this attitude the bird hops rapidly along the branch, calling *wa-wa-wa-wa* (the "pump call") as he bounces up and down, the movements of his body increasing the splendor of the cascade of plumes above and behind him. Omitting the first phase, the bird may start with this display as soon as a female arrives at his station. A single performance lasts at most ten seconds, but it may be repeated several times.

The second phase leads to the third, "the bow," in which the bird's body is gracefully arched, with his head and tail below the perch, his wings spread around it as though in an embrace, and his plumes raised above his back. This lovely attitude, which the bird holds rigidly from a few seconds to more than a minute, was more poetically called by Gilliard the "flower pose." A bird might take this pose in the absence of a female, but it was always assumed when a female arrived.

Phase 4, which Dinsmore called "the dance," is perhaps better called "hopping," since all the displays together are often designated a dance. Crouching low, the actor moves slowly and rhythmically back and forth along the branch, bouncing with both feet simultaneously in the air. While hopping, he repeats at intervals of about one second a *click,* such as a person can make with the tongue against the roof of the mouth. After hopping from a few seconds to more than a minute, the bird usually wipes his bill several times against his perch, shakes his wings, and lowers his plumes.

The culminating phase is "mounting." The male bird of paradise rubs his bill against the female's bill, then stretches one wing protectingly over her and holds her close to himself while he flaps both wings. Extending his head and neck beneath hers, he rubs his bill against the far side of her head while he continues to hop. Finally, still flapping his wings, the male mounts the female. After brief union, she flies away, while he continues for a few minutes more to display, call, and preen on his dance limb.

The most noteworthy differences in the performances of the several species of *Paradisaea* appear to be in their climax, or flower, displays. Instead of holding his side plumes as erect as the Greater bird does, the smaller, daintier Lesser, with body gracefully arched and wings widely and downwardly spread, throws upward his curving, lacy plumes, shakes them vigorously to separate the filaments, then permits them to flow down behind his tail in a dense, shimmering, white-and-yellow cascade that captivates the eye. Count Raggi's Bird of Paradise bows so far forward on his perch that his body is almost inverted. His wings are widely spread to form a semicircular curtain in front of his head; his red plumes rise in splendor high above him. The plumes of the Emperor of Germany Bird of Paradise are much shorter and looser, almost as fine as cobweb. After a short, simple dance, he drops below his perch to hang upside down, suspended by his feet, while he sways and twists to make his filmy plumes swirl over his underparts in a gauzy tangle, through which shine his metallic green throat and breast.

The small Prince Rudolph's, or Blue, Bird of Paradise, whom admirers have called the most gorgeous member of his genus, also has a pendent display, which C. R. Stonor, who watched it in an aviary, described as follows:

> In common with its relatives, it has a loud and penetrating call, again of an individual nature, and once again ushering in the display. The call is followed by a curious low, grating song; as the bird croons away to itself, it sinks gradually lower and lower on its perch, eyes half-shut, and oblivious to all around it. A few seconds later it is upside-down, hanging vertically, and in full display of its plumes. Brilliant though it is in rest, in display it is almost beyond words. The violet-blue plumes are spread out over the underside of the body in a fan, so that the patches on each side merge one into the other; it holds its head almost straight down, so that the apex of the fan is on the throat . . . The plumes are moved by slight and most dextrous manipulations, so that shimmering waves of blue and violet pass across the fan, while the red and black oval [at the center of the fan] is now broadened and now narrowed, as the feathers making it up are brought together or pulled apart. To complete the picture, two of the tail feathers, greatly lengthened to long narrow strips like dark-blue ribbons, droop down on either side of the fan as the bird hangs suspended.
>
> All the while their owner sways to and fro, keeping up his low grating song in a curious rhythm, more or less in time with his swaying movement. As suddenly as it was unfolded, the fan is shut, the song ceases, and the bird goes back to normal.

Although the upside-down display is most fully developed in the two foregoing species, it is given occasionally by several other birds of paradise, including the Greater, the Lesser, and the Red. To assume the vertical pose, the birds may fall over either forward or backward. An inverted display is also given by the very different, insectivorous Buff-tailed Sicklebill (*Epimachus albertisi*). Inverted displays are of critical importance for the question of whether adornments and displays were developed primarily to intimidate rival males or to impress females. A bird trying to dominate a rival would hardly hang upside down in front of him, in a submissive rather than an aggressive attitude; but a male trying to charm a female might assume such a posture the better to reveal his beauty. Inverted displays are more compatible with intersexual than with intrasexual selection.

Birds of paradise whose adornments are quite different from those of *Paradisaea* have, of course, very different displays. Those of the Magnificent and the King of Saxony were described earlier in this chapter. To tell about all that are known would be tedious; the more splendid they are, the less adequate verbal descriptions become. Nevertheless, in an effort to do justice to this resplendent family, I shall describe two more.

The three species of riflebirds are true birds of paradise that received their puzzling name because they are black and green, like the uniform of riflemen in the British army of the early nineteenth century. The thirteen-inch (33-centimeter) Magnificent Riflebird (*Ptiloris magnificus*), widespread at low altitudes in New Guinea and northeastern Australia, has a long, slender, curved bill and a short tail. The prevailing velvety blackness of the male's plumage is relieved by his glossy green crown, throat, chest, and central rectrices. At first sight, he does not appear to have any special adornment to display, but he springs a surprise. More gifted vocally than most birds of paradise, he has a clear, melodious whistle, rising in pitch and thrice repeated, that rings through the dense forests where, shy and wary, he remains so well hidden in the treetops that for a full description of his display we must depend upon observations made in aviaries. Here, after uttering his stirring call, he flits excitedly from perch to perch, then suddenly stands and stretches upward to his full height. With a rushing sound and a flash of black, he spreads his wings to their fullest extent, making a pattern that one never expects of a bird. With no projecting angles and not a gap in the expanse of dark plumage formed by his wings and body, the neatly rounded figure resembles a thick crescent, with the bird's head and breast projecting into the concavity at the top. First, he lays his head to one side, thrown back to give prominence to his shining green gorget. Then he begins to move his head across the concavity, from one wing to the other, at first slowly, then back and forth with increasing speed, until only flashes of green are discernible. Simultaneously with his head movements, he waves his wings up, down, and forward, with a swishing sound made by the rough edges of the feathers as they rub together. Flicking his head, arching his

Magnificent Riflebird, *Ptiloris magnificus,* male displaying.
New Guinea and northeastern Australia.

wings, the Magnificent Riflebird dances gracefully back and forth along a branch.

The Superb Bird of Paradise (*Lophorina superba*) is a thrush-sized species of mid-altitude forests through the length of New Guinea. The male is black with a green crown. Covering his back like an umbrella is an erectile cape composed of long, velvety black plumes that spring mostly in tufts from each side of his hindhead. On his breast is a patch of elongated, iridescent green feathers, longest at the sides. When about to display, he utters a harsh, piercing screech, unworthy of so splendid a creature. Then he starts hopping to and fro along a branch, screeching louder and louder. Suddenly, his folded cape springs upward and spreads widely, surrounding his head like an oversized halo. Simultaneously, the shield on his chest expands broadly. The pointed ends of this shiny pectoral shield almost meet the tips of the bronzy-black halo. In the midst of this wide oval expanse of shimmering plumes rises the bird's green-capped head, with his mouth wide open to display its bright apple-green lining, and in front of each eye a tiny eyelike patch of scintillating green erectile feathers that mask the real eye. As he trips along his perch, the strangely transformed bird calls attention to himself by occasional screeches and sharp, clicking sounds made by snapping his wings against his sides. After two or three minutes of this fantastic performance, the halo, the pectoral shield, and the eye tufts fold together in a trice; the Superb Bird of Paradise becomes again a bird who can fly and forage. In contrast to that of many other birds of paradise, the diet of the Superb contains only about 25 percent fruits, which it gathers on territories dispersed through the forests.

When we contemplate the forty-three species of birds of paradise, we are impressed no less by the diversity than by the splendor of the males' adornments and displays. Their ornamental plumes may spring from almost any part of the body: the head, neck, chest, sides, wings, or tail, and they are exceedingly diverse in form. The family reveals a strong tendency to produce ornamental feathers but no consistent trend in their position or shape. How can we account for this strange situation?

I believe we must postulate, in the ancestral birds of paradise, a degree of mutability of plumage, especially in the males, unusual among birds. However, this alone does not appear adequate for the evolution of the extremely elaborate ornamentation of many of these birds. Mutations tend to be random; for large effects, they must be supplemented by selection in a determinate direction. These adornments hardly increase the birds' efficiency in finding food, escaping predators, or keeping warm; they have no obvious function apart from their nuptial displays. The agents of selection can be no other than the females to whom they display. Therefore, in addition to extraordinary mutability in the plumage of the males, it appears necessary to postulate, in the females, an exceptional sensitivity to visual impressions, and a persistent preference for the more beautiful or the more spectacular display. Sensitive to visual beauty, they appear to care little for auditory beauty. The aesthetic sense need not be equally developed in all directions. Just as some people prefer beautiful sights to beautiful sounds, while others delight more in sounds than in sights; so some birds appear to be more strongly impressed by appearance than by voice. This may explain why birds of paradise, resplendent in plumage, have mostly unmelodious calls, whereas other birds, plainly attired, sing enchantingly.

References

Beehler 1983a; 1983b; 1987a; 1987b; 1988; Beehler and Pruett-Jones 1983; Dinsmore 1970; Gilliard 1969; LeCroy 1981; LeCroy, Kulupi, and Peckover 1980; Pruett-Jones and Pruett-Jones 1988a; 1988b; Rand 1938; 1940; Stonor 1940; Wallace 1872.

11.
The Courtship of Bowerbirds

The Australasian region, which includes Australia, New Guinea, and neighboring islands, is the home of two families of birds without counterparts elsewhere, not even in the Neotropical region, which has a much greater number of avian species. One of these unique families is the megapodes or moundbirds, the only birds that, instead of incubating their eggs with their own bodies, depend wholly upon different sources of heat, sometimes in carefully controlled incubation mounds. The other unique family is the bowerbirds, most of which build special, often elaborate structures for courtship alone. Manakins of the New World prepare stages for courtship, by clearing terrestrial courts or stripping leaves from branches, but they do not build anything for this purpose. Males of a number of other families start or complete nests to entice females, but such nests are commonly used for eggs and young. Many other birds build nests for sleeping. Bowerbirds alone make structures quite different from nests and bring ornaments to them. Not even birds of paradise impart such uniqueness to the Australasian avifauna; although the most lavishly adorned of avian families, they are not the only birds with splendid ornamental plumes.

The eighteen species of bowerbirds range in size from 8.5 to 15 inches (22 to 38 centimeters). In about half the species, both sexes wear cryptic green, gray, or brown. In the remainder, males are resplendent in yellow, orange, and black, they wear high yellow crests, or they are shiny blue-black, while females are more plainly attired. Bowerbirds are found at all altitudes up to timberline, mostly in wet forests but also in more open arid country and grasslands with scattered trees. Largely frugivorous, they diversify their diets with insects, other invertebrates, and occasional small vertebrates. In open cups in trees or vine tangles, the female lays one or two, rarely three, eggs, which are plain whitish, buffy, or greenish, or variously blotched and scrawled. For about nineteen to twenty-four days she incubates without help. During a nestling period of eighteen to twenty-one days, she attends the young, also without a mate's assistance, except among the catbirds.

The three species of catbirds of the genus *Ailuroedus* differ greatly from all other members of the bowerbird family. The sexes scarcely differ in their largely green, spotted and streaked plumage. The only birds of their family known to form pairs, they build no bowers but, as though to compensate for this omission, they make, of twigs, vines, and many large leaves, nests that are bulkier and more substantial than those of the bower-builders. Unlike all of the latter, so far as known, male catbirds attend their young.

The Stagemakers

The ten-inch (25-centimeter) Tooth-billed, or Stagemaker, Bowerbird (*Scenopoeetes dentirostris*) occupies a restricted range between two thousand and five thousand feet (610 and 1,525 meters) in the rain forests of northern Queensland, Australia. Both sexes are brownish olive above, with pale underparts streaked with brown. Their food includes fruits, snails whose shells they break on an "anvil" beside the display ground, and leaves that they pluck with strong, toothed bills. Loud voices reveal the locations of the males' courts amid the dense undergrowth of the forest. Like manakins of the genus *Manacus,* six-wired birds of paradise, and Magnificent Birds of Paradise, the Stagemaker clears an oval or circular patch of ground by tossing aside or carrying away all fallen leaves and other removable debris. The area, which appears to have been swept clean with a broom, may be as small as four by five feet (1.2 by 1.5 meters) or as large as seven by ten feet (2.1 by 3 meters). Often a few slender saplings grow in its midst, and a fairly large trunk stands in or beside it.

Unlike the above-mentioned manakins and birds of paradise, the Stagemaker does not permit his court to remain bare but adorns it with large leaves, whose petioles he severs from living plants by laboriously chewing or sawing with his toothed bill. Some of these leaves are twice his own length. When he reaches his stage with them, he nearly always lays them upside down, so that the paler undersides contrast more strongly with the dark ground upon which they lie. The birds are known

Tooth-billed Bowerbird, *Scenopoeetes dentirostris,* male arranging leaves on his court. Northern Queensland, Australia.

to select at least seventeen species of leaves, but they have individual preferences. When A. J. Marshall replaced certain kinds of leaves with an equal number of others, the owners of the courts removed the alien leaves and brought more of the kind they favored. When the leaves wither, the Stagemaker casts them aside and brings fresh ones.

The Stagemaker's visual displays are poorly known. John Warham watched a male emerge from behind a trunk and, in a crouched attitude, hop jerkily over his court, flicking his wings outward and his tail upward, "uttering typical bower-bird cacklings and hissings." No other bird was in view, and the courtship of a Stagemaker in the presence of a female was not seen. His vocal performances have frequently been described. On a perch above or beside his court, from ground level to about twenty feet (6 meters) up, a Stagemaker pours forth his varied notes for long intervals. Some observers credit him with being an excellent mimic; others disagree. In any case, his medley includes both melodious and harsh notes, some so powerful that they make a human ear throb. At the opposite extreme, he whistles softly and continuously. He spends much more time aloft than on his ground court, on branches worn smooth by his feet.

Stagemakers' courts tend to be associated in exploded leks, situated along wooded ridges, with one male sometimes within hearing of four or five others, whom he cannot see because of dense intervening undergrowth. Warham noticed that during pauses in his singing a Stagemaker listens to the voice of a neighbor, and may turn to face him when he resumes his recital. The naturalist repeatedly noticed that a phrase was tossed back and forth between two singers, who continued to repeat it until one of them introduced a new verse, which was promptly imitated by the other. The songs of these birds differed from day to day. They must serve not only to keep rival males in touch but also to guide females to the stages. As she approaches, she may be helped by a visual clue, for while a male sings, the movements of his mouth rearrange his throat feathers, revealing their paler bases and making the bird easier to detect in the somber undergrowth. Only females attend their flimsy, saucer-shaped nests of twigs, built from fifteen to a hundred feet (4.6 to 30 meters) up in the forest, each with two creamy brown eggs.

A further elaboration of the ground court is made by Archbold's Bowerbird (*Archboldia papuensis*), a fifteen-inch (38-centimeter) dark gray to black bird with or without a golden crest. These inhabitants of New Guinea's highland forests between 6,700 and 12,000 feet (2,040 and 3,660 meters) were not discovered until 1939. High on Mount Hagen, in montane forest with a moderately dense undergrowth of shrubs, pandanus, and tree ferns, the indefatigable ornithologist-explorer Thomas Gilliard found five bowers of this rare bird within about two miles. Three were in use and two abandoned. In diameter they ranged from three to eight feet (0.9 to 2.4 meters). The central feature of a display area was a plot of cleared ground that the owner had covered with a tangled mat of ferns and vines, surrounded by ground ferns and golden-yellow strands of climbing bamboo draped over low branches by the bird. Lying on the mat were pieces of charcoal, a cluster of unbroken black snail shells, a pile of broken gray snail shells, an accumulation of beetle wings, a cluster of black and amber chips of resin, and a little mound of seeds. Close by the main display area were three small spaces apparently cleared by the owner of the bower.

Entering his blind early on the morning of July 14, 1956, Gilliard heard Archbold's Bowerbird utter a variety of notes ranging from insect-like buzzing to earsplitting whistles. At intervals the dusky bird descended to the bower's floor to shift and play with his display things, and he brought a long tendril or strip of some other vegetation. Finally, in midmorning, a female arrived and incited the strangest of courtship displays. Not only did the roles of the sexes seem to be reversed (as in phalaropes, jacanas, and certain other birds), but the male appeared to be utterly cowed and submissive to his black visitor. He dropped to his mat of ferns and lay flat upon it, his crest folded back against his head, his body, half-open wings, and partly spread tail pressed against the mat, while he continued to utter a low *churr purr, purr churr, churr.*

While the male lay in this abject attitude, so flattened that he resembled a reptile more than a bird, the female flew across the court from perch to perch at its sides. As she passed close above him, she whipped her wings so rapidly that they sounded as though they would be torn. This wing-beating, with a ripping sound like cardboard being torn, was her only display toward the male. Whenever she alighted, he turned and crept very slowly toward her, "like a whipped dog crawling toward its master." Occasionally, the male interrupted his creeping to hop forward a few inches. Part of the time he held a thin strand of bamboo or fern in his bill. The female would delay on her perch until he advanced to within a foot of her before she flew over him, often hovering to whip her wings above his back, to alight on another side of the bower. After the male had groveled before her for more than twenty-two minutes, she flew away without descending to his mat of ferns; he followed. Of all the birds whose courtship is described in this book, no other female has appeared so aggressive; mostly females who visit displaying males are passive onlookers. No other male has appeared so abject. The nest of Archbold's Bowerbird was unknown to Gilliard.

The Avenue-builders

Bowerbirds of three genera are known as avenue-builders because of the form of their constructions. They begin by covering a small plot of ground with a thick mat of sticks crossing in all directions. Into this platform of sticks they insert upright twigs in two parallel rows, a few inches apart. The space between these walls is the avenue.

One of the most colorful of the bowerbirds is the Golden Regent or Flame Bowerbird (*Sericulus aureus*), resplendent in orange and yellow, with (in one race) a black face. An avenue found by Gilliard in dense forest on the Tamrau Mountains of western New Guinea was only seven inches (18 centimeters) long, between walls no more than ten inches (25 centimeters) high. On the floor of the avenue were five blue berries and a black shelf-fungus about one inch wide. The ground at each end of the avenue had been cleared but was apparently without decorations. The meagerness of the bower and its adornments suggests that, more than other avenue-builders, the Golden Regent depends upon his colorful plumage to impress females. The elegant, ten-inch (25-centimeter) black-and-golden yellow Australian Regent Bowerbird (*S. chrysocephalus*), restricted to the extreme east of that continent, also builds a small bower, in lowland rain forest. With macerated, pea-green vegetable material, held in his bill and mixed with saliva, he paints the sticks in his bower's walls. On the floor of the avenue he deposits blue, red, and black berries, shells of land snails, pinkish young leaves, or yellow flowers.

Best-known member of the family is the twelve-inch (30-centimeter) Satin Bowerbird (*Ptilonorhynchus violaceus*), which inhabits the woodlands of eastern Australia and enters gardens in the suburbs of its largest cities. The male's satiny black plumage glistens with tints of violet, purple, and blue. The dull green female has dark crescentic marks on her creamy yellow underparts. Both sexes have bright blue eyes. Like other bowerbirds, they eat chiefly berries and other fruits, with an admixture of insects, the main food of nestlings. After the breeding season, some gather in flocks and wander widely; others remain near their bowers.

Beneath trees, in a cleared space three or four feet long by twenty to thirty inches wide (90 to 120 by 50 to 76 centimeters), Satin males build their bowers. The parallel walls, composed of dozens of thin twigs, are about twelve inches (30 centimeters) high, three or four inches (8 to 10 centimeters) thick, and slightly shorter than they are tall. They arch over the avenue between them, which is four or five inches (10 to 13 centimeters) wide. This central passage usually runs approximately north and south. When A. J. Marshall experimentally shifted a bower to another orientation, the owner altered the walls until his structure regained its original direction. The advantage of this north-south orientation appears to be that in the early morning, when through the avenue a female watches the male performing on his display platform in front of it, each can watch the other without staring into the bright rays of the rising Sun. An adult male can complete a bower in a day or two. With repeated reconstruction, a bower, or at least its site, may be occupied by the same male for fifteen years. Males build their bowers one hundred yards or more apart.

The platform of crisscrossing sticks on which the avenue is erected is always longer than the avenue itself. On the open part of the platform, at one end of the avenue, the male bowerbird accumulates his treasures, revealing a marked preference for those that are blue (the color of the eyes of both sexes) and yellowish green (the color of the plumage of

Australian Regent Bowerbird, *Sericulus chrysocephalus,* male.
New South Wales to southern Queensland.

female and immature male Satin Bowerbirds). Other acceptable colors are pure yellow, brown, and gray. The objects of these colors that the birds bring to their bowers are extremely varied, including flowers of many kinds, blue parrot feathers, brown land shells and cicada exuviae, and fungi. Near human habitations, the birds accumulate blue plastic artifacts, fragments of blue glass and pottery, scraps of blue paper and rags, laundry bluing, and similar oddments, the whole forming a colorful display in the dark woodland undergrowth. Adult males are not above pilfering attractive baubles from the bowers of neighbors, who, when opportunity permits, retrieve them, or steal from the thieves. Even in the mating season, Satin Bowerbirds travel long distances, and have been known to carry an object from one bower to another two miles (3 kilometers) away. On occasion, an adult male wrecks a bower built too near his own, always furtively, in the proprietor's absence, ready to flee the moment he reappears. Gerald Borgia learned that the better the construction of the bower, and the more abundant its adornments, the better its chances of escaping destruction by other males.

Many, but not all, adult male Satin Bowerbirds paint the walls of their avenues. Some use for pigment the pulp of berries, others prefer charcoal or green liverworts, and, when provided for them, laundry bluing is eagerly accepted for coloring. The painter mashes or grinds the material in his beak and mixes it with his saliva. With fibrous bark, he forms a small, spongelike wad to retain his mixture in his partly open mouth while, with the side of his bill, he applies it to the sticks. Although he does not employ the wad as a brush, this may be considered as one of the few known examples of tool-using by birds. Since the thin, sticky black, blue, or green coating washes off the sticks in a moderately hard rain, the Satin bird replaces it frequently in the season of his most active display.

Male Satin Bowerbirds need from four to seven years to acquire the glossy blue-black plumage of full maturity. During at least the later years of their nonage, they, too, build bowers that are often smaller and less substantial than those of older males, and they bring colorful trinkets to their construction, which may be shared by half a dozen of them. Early in the season, a mature male permits these youngsters to build and display in his territory, but later he destroys their rudimentary bowers and carries to his own any decorations that appeal to him. By practicing the art of bower-building, visiting bowers of mature males while the owners are absent, and watching adults display, young birds improve their own constructions and displays.

Mixed flocks of Satin Bowerbirds reveal their presence by a chorus of croaking, explosive sounds and whirring rattles. When displaying at their courts, males chatter, buzz, creak, and utter ringing cries. This harsh outburst is followed by a bout of vocal mimicry, during which the birds imitate postman's whistles, cat cries, and calls of other birds. Frequent models are the Laughing Kookaburra (*Dacelo gigas*) and Lewin's Honeyeater (*Meliphaga lewinii*). Just as male Satin birds take years to acquire full adult plumage and build perfect bowers, so they take long to become accomplished mimics. Christopher A. Loffredo and Gerald Borgia collected evidence that females prefer to mate with older males who are the best mimics. The harsh or mechanical part of the males' vocal outpourings differs less between individuals and, apparently, has less influence upon the females' choices. They also prefer males whose bowers are well made and profusely decorated. By monitoring bowers with automatic movie cameras, Borgia demonstrated great differences in the number of matings by individual males. Those with the most snail shells, blue feathers, and yellow leaves tended to attract the most females. In a sample of twenty-two males, five accounted for 58 percent of all matings.

When a female Satin Bowerbird visits a male's bower, he becomes so excited that his eyes become rose-red and seem about to pop out of his head. Voicing a medley of scolding and churring notes, unpleasant to the human ear, he picks up a berry, snail shell, or twig and holds it in his bill while he hops and jumps about erratically, often encircling the avenue. He flings his wings outward in a colorful gesture. While he performs on the open platform at one end of the bower, the demure female at the other end watches him intently through the avenue. Sometimes she chatters much as he does. Occasionally, she pecks at the bower's basketwork or picks up leaves or other display things, only to drop them. She may rearrange sticks in the wall, even lift up and add a new one, or go through the motions of painting. If she ventures to the front of the avenue, the male may chase her mildly around and around its walls. He appears to be in a highly ambivalent state, resentful of this intrusion into his zealously guarded bower, yet driven to accept and court his visitor. Her close resemblance to the immature males, whom he tries to exclude from his domain, doubtless intensifies his inner conflict. Often his vehemence frightens the female prematurely away.

If the female bowerbird does not timidly depart and intruding males do not interrupt courtship, the male mounts her on the bower, usually in the avenue. His violently beating wings may partly wreck his walls. After a courtship that may be prolonged for half an hour, followed by coition, the female appears so exhausted that she can hardly drag herself from the bower to recuperate in the shelter of nearby bushes. If she stays too close, the male may churlishly drive her away. Then, taking no further notice of her, he proceeds to repair his damaged avenue, and to preen. Although early observers surmised that Satin Bowerbirds might be monogamous, Sid and Reta Vellenga's study of banded birds showed that in one season a single male mated with five females and courted many others. No female that he had inseminated was seen to return in the same year, and only one mated with this male in two consecutive years. Later, Borgia's cameras filmed thirty-three matings, with an unstated number of different females, by the most favored male in one season. In a shallow open bowl of twigs, lined with leaves, the female lays one to three, usually two, dark cream-colored, spotted eggs. She hatches them and rears her young with no male's help.

The male Satin Bowerbird's treatment of a visiting female is an instructive example of the difficulty of harmoniously integrating blind instinctive drives with more deliberate behavior in an animal with dawning intelligence, such as the bowerbird appears to be. Apparently, he becomes strongly attached to his bower but does not understand its ultimate purpose. When a female arrives to enable him to fulfill this purpose, her intrusion into his cherished precinct upsets him, but he is driven by a powerful impulse to accept and court her. The clash of motives causes highly irrational behavior. Humans, for all their mental development, are still beset by such distressing conflicts between instinctive drives and rational conduct.

Over vast stretches of Australia's arid interior, the Spotted Bowerbird (*Chlamydera maculata*) replaces the Satin Bowerbird of the continent's rainier eastern fringe. The sexes of this eleven-inch (28-centimeter) bird differ little in their brownish plumage, boldly spotted on back and wings with rufous and golden buff. On the back of the neck is a small, erectile frill of long, loose-barbed rose-lilac feathers, which on some females is reduced or lacking. Clumps of bushes and trees scattered through open country are chosen by males as sites for their bowers, with preference for trees whose branches almost touch the ground and offer good concealment. In such places, never far from water, they build avenues much like those of Satin Bowerbirds but often larger. Near the center of a display ground about six feet (2 meters) long, a male erects walls ten to twenty inches (25 to 50 centimeters) high and fifteen to thirty inches (38 to 76 centimeters) long. These walls, composed of thin twigs on the outside and grass stems on the inner face, are from five to nine inches (13 to 23 centimeters) thick. They enclose an avenue six to nine inches (15 to 23 centimeters) wide.

The Spotted Bowerbird is a tireless collector of trinkets to spread over

Spotted Bowerbird, *Chlamydera maculata*. Sexes similar.
Arid inland Australia.

his platform of sticks at both ends of his avenue. Since the introduction of rabbits and sheep to the southern continent, their bleached bones, especially the whitened vertebrae of the latter, have become favorite adornments. Well over a thousand have been counted on a single bower. Bivalve shells and water-worn stones, apparently brought from rather distant streams, add to the hoard. Green seedpods, berries, and seeds diversify the collection. Where available, fragments of glass and pieces of hardware — nails, screws, bolts, bits of wire, brass cartridge cases — lie amid the bleached bones. This bird has acquired an unenviable reputation as a thief who boldly enters homes and camps to carry off scissors, knives, forks, spoons, thimbles, coins, and jewelry. He has even pilfered the ignition keys of a parked car! Fortunately, he does not try to hide his stolen goods; those familiar with his ways can often retrieve what they have lost at the nearest Spotted bird's bower. In strange contrast to his predilection for shiny metals and fragments of glass, he usually disdains bright greens, reds, yellows, or blues — just the colors often preferred by bowerbirds of humid regions. Unlike the Satin bird, he strews his display things along the length of his avenue. He paints the grass stems on the inner sides of its walls with a stain made by chewing dried grass and mixing the extract with saliva.

The Spotted Bowerbird collects sounds as assiduously as he collects solid objects. Marshall wrote that he is "probably the most gifted mocking-bird known." In addition to mimicking with great fidelity the voices of many of the birds that surround him, he can reproduce such diverse noises as those made by an Emu crashing through twanging fence wires, cattle breaking through scrub, and a maul striking a wood-splitter's wedge. Yet, despite his vocal virtuosity, his own proper notes appear seldom to be melodious. He utters a ringing, somewhat metallic advertising call and, when displaying, harsh scolding and hissing notes that would seem to repel rather than blandish his female visitors.

The Spotted Bowerbird's wooing, like that of the Satin bird, is stormy. To the accompaniment of loud hissing, scolding, and rhythmical mechanical sounds, he postures in various strained attitudes on his platform at either end of the avenue. With one of his trinkets in his bill, he shows his expanded rose-lilac crest to his visitor, while facing the avenue entrance or while standing transversely to it and bending his head sideward, in either case jerking it up and down. He flings his bone or shell from his beak as though angry with it. He leaps his own height into the air. At intervals he leaves his platform and, with tail cocked up and wings loosely drooping, circles the bower, walking or running around it instead of hopping, his usual gait. In his great excitement, he sometimes trips and falls. Appearing timid, the female watches the violent courtship silently, or at most with a little hiss. She is careful to keep the walls between herself and her wooer. After prolonged demonstrations, mating may occur outside the bower.

The biggest member of its family, the Great Gray Bowerbird (*Chlamydera nuchalis*) is fourteen to fifteen inches (36 to 38 centimeters) long. The brownish feathers of the adult male's back and wings are tipped with ashy gray, giving him a mottled appearance. He wears an erectile, rose-lilac crest, much like that of the Spotted Bowerbird. Below, he is pale gray. The slightly smaller female is paler and often lacks the nuchal crest. The species is widely distributed across Australia's far north, in tropical scrubland more humid than that inhabited by the Spotted bird.

The Great Gray bird's bower, similar to that of the Spotted bird but often larger, is usually oriented with the avenue's long axis running nearly north and south. The side walls slope inward until they almost meet, partly roofing the passageway. John Warham watched males, their mandibles glistening with saliva, wipe their bills up and down the twigs

Color plate 4 (on following page).

Top, left to right: Wilson's Bird of Paradise, *Diphyllodes respublica*, male, small islands west of New Guinea; King Bird of Paradise, *Cicinnurus regius*, male, New Guinea and neighboring islands. Bottom: Blue Bird of Paradise, *Paradisaea rudolphi*, female, above, and male in inverted display, mountains of eastern New Guinea.

of these walls, but their painting movements appeared to have little effect. An assiduous hoarder, the Great Gray Bowerbird deposits his collections inside the avenue as well as on the platform at both ends. His acquisitiveness appears not to be restrained by good taste. Like his Spotted relative, he has a predilection for pale objects, such as the bleached bones of kangaroos and wallabys, shells, and pebbles of quartz, limestone, or laterite, hundreds of which may litter his platform. Fresh green leaves, green berries, green plant galls, or pale green flowers may diversify his collections, which near human dwellings may contain any small, glittering object that he can carry off, sometimes from inside houses that he boldly enters. Aside from green, these birds appear to disdain pure colors, although some select red artifacts. While they continually shift their treasures from one pile to another, they often drop them with a clink or metallic tinkle that reveals the location of their bowers. The birds seem to enjoy these sounds. Their displays to visiting females, in which presentation of the nuchal crest is a prominent feature, differ little from those of Spotted Bowerbirds. Great Gray Bowerbirds of both sexes are also accomplished mimics.

The sexes of the eleven-inch (28-centimeter) Fawn-breasted Bowerbird (*Chlamydera cerviniventris*) hardly differ. Above, both are dusky brown, with pale spots; below, they are tawny buff, streaked with dark brown on the breast. They lack the lilac crest of the two preceding species. Widely distributed through the coastal lowlands of New Guinea, where they occasionally ascend as high as 3,600 feet (1,100 meters), Fawn-breasted Bowerbirds range through the islands of Torres Strait to the northeastern top of Australia. Their preferred habitat is grassland with scattered clumps of trees and fringing low, open woods. The avenues of these bowerbirds are of the same form as those that we have already met, but they are built by inserting upright twigs into platforms of interlaced sticks that are often much thicker, sometimes as much as fourteen inches (36 centimeters) high. On poorly drained ground subject to torrential downpours, this may save the avenue and display objects from flooding. The latter consist almost wholly of green fruits of various sizes, detached or united in sprays, which the male bird not only distributes over the platform at both ends of his avenue and along its length but also hangs precariously on top of the walls above it. When these green ornaments decay, he removes them to a refuse heap beyond the bower. Rarely he brings a few bleached bones or shells to his bower.

Fawn-breasted Bowerbirds imitate the ventriloquial notes of neighboring species of birds and have a wide repertoire of unmelodious sounds, churring, hissing, sputtering, rasping, rattling, and whining. Ut-

Avenue bower of Great Gray Bowerbird, *Chlamydera nuchalis.*
Northern Australia.

tering some of these harsh or explosive noises, a male displays to a female crouched in his bower while he holds a cluster of green berries or
a green palm seed in his bill and vigorously jerks it up and down. He
twists his neck as though to display to the watching female a crest on his
hindhead that he lacks. The similarity of this posture to that adopted by
Spotted and Great Gray bowerbirds to show their rose-lilac crests led
Gilliard to postulate that the Fawn-breasted's ancestors had similar
crests, which became superfluous after they performed with a green

berry in their bills, and in consequence were lost in the course of the bird's evolution — an example of Gilliard's "transferral effect."

Both sexes of the eleven-inch (28-centimeter) Yellow-breasted, or Lauterbach's, Bowerbird (*Chlamydera lauterbachi*) have olive to brown upperparts and yellow underparts, with dark stripes on throat and chest. They are crestless. At low and middle elevations in New Guinea, they seek fruits and insects in grassy fields with scattered shrubs and small trees or just within the edges of adjoining woods, usually near marshy ground or running water. Their bowers are more complex than those of the other three species of *Chlamydera*. On the usual foundation mat of interlaced horizontal twigs, the male inserts sticks to form a structure with four walls instead of two. The walls enclosing the main avenue slope outward instead of standing upright and arching over the passageway. The inner faces of these walls of sticks are covered, much as a nest is lined, with long, fine strands of brown grass, some of which shade the avenue. Opposite each end of the avenue the bird builds a transverse wall. The four walls are arranged like the pen strokes of the equation $1 = 1$. From no exterior point is it possible to look down the avenue of this construction with four entrances. Sixteen bowers measured and weighed by Gilliard ranged from 28 to 38 inches (71 to 97 centimeters) in length, from 19 to 26 inches (48 to 66 centimeters) in width, and from 14 to 25 inches (36 to 64 centimeters) in height. The central passageway was about 7 to 12 inches (18 to 30 centimeters) long and 2.25 to 3.25 inches (5.7 to 8.3 centimeters) wide. These structures were so strongly built that they remained intact when lifted. They weighed from 6.5 to 16.5 pounds (3 to 7.5 kilograms).

The heaviest of these bowers was made of over three thousand sticks and was lined with more than one thousand hairlike strands of brownish grass. Its weight was due mainly to nearly a thousand small, pale slate-colored pebbles, its most abundant ornaments. These were placed in the center of the main avenue and in the transverse passages at its ends. In some bowers the tiny stones were inserted among the sticks on the inner sides of the steeply sloping end walls, facing the central avenue. Often they were stuck between the sticks with such masonlike precision that they formed a miniature wall several inches high and wide. Although other avenue-builders accumulate pebbles, they are simply piled up or strewn about, not inserted into the walls. Other ornaments of the Yellow-breasted Bowerbirds are berries, red, blue, and green. All these display things are placed within the four walls; the foundation mat, over which other avenue-builders scatter most or all of the adornments, hardly extends beyond the walls of the Yellow-breasted Bowerbirds' constructions. These birds appear not to paint the grass-lined walls of their avenue.

Gilliard watched a male Yellow-breasted bird with a red berry in his bill display to a female who entered his bower. These bowerbirds without sexual differences in plumage, plainly attired and crestless, but with the most elaborate constructions of all the avenue-builders, provide additional evidence for the transferral effect.

The Golden Bowerbird

The Golden Bowerbird, also called the Queensland Gardener (*Priono-dura newtoniana*), smallest member of its family in Australia, builds the largest bowers. Slightly over nine inches (23 centimeters) long, the male is golden olive with yellow or orange-yellow crown, hindneck, outer tail feathers, and underparts. The female is olive above and ashy gray below. This beautiful bird dwells in the rain forests of northern Queensland from about 1,500 to 5,400 feet (460 to 1,645 meters) above sea level. Berries and other small fruits are its principal foods. For his bower, the male chooses two slender saplings growing upright close together. Around the base of each he piles an immense number of twigs in a roughly pyramidal mass, which on the preferred sapling may rise to a height of nine feet (2.7 meters) but is usually much lower. The pile of twigs around the second sapling is typically much less tall than that about the first. These two very unequal piers are always joined by a sort of bridge, formed by one or more horizontal vines or branches that the builder found already present. The Golden bird keeps the center of this bridge bare of twigs and displays upon it. The sticks in this construction are stuck together with a fungal growth, as are those of the female's nest. The bower is decorated, chiefly in the vicinity of the display perch, with lichens, mosses, ferns, flowers, fruits, and seeds, brought at intervals by the owner during the season of display. Scattered around some of these bowers are a number of tiny structures made of twigs piled to a height of about eighteen inches (46 centimeters), like the first stages in the construction of a principal bower. They resemble miniature aboriginal huts, and together give the impression of a native camp.

An accomplished mimic, the Golden bird imitates the calls of the Stagemaker, Green Catbird (*Ailuroedus crassirostris*), Queen Victoria Riflebird (*Ptiloris victoriae*) and others. Most often heard at his bower are the harsh notes typical of bowerbirds: croaks, pulsating buzzes, hisses, rattles, mechanical sounds, and "scolding" notes. If an intruder misplaces one of his ornaments, the careful bird returns it to its proper place at his earliest opportunity. I have found no account of the Golden bird's display to a visiting female.

The Hut-builders and Garden-makers

In the gardener bowerbirds of the genus *Amblyornis* we can trace the elaboration of the bower to its highest artistic expression, while the bower's maker becomes increasingly drab. Without his long, full, yellowish orange crest that extends far behind his head, the ten-inch (25-centimeter), olive-brown Macgregor's Gardener Bowerbird (*A. macgregoriae*) would be as severely plain as the crestless olive-brown female. On forested ridgetops in New Guinea's rainy mountains, the males build structures of the "maypole" type, very different from that of the Golden Bowerbird. Around a slender, upright sapling, the bowerbird lays small sticks horizontally, to build up a bristly column two or three feet (60 to 90 centimeters) high by about half as thick at its widest part. Around the base of this tower he makes a circular runway of gray-green moss, so well compacted that it can be rolled up like a thick carpet. At the outer edge he builds up the moss into a rim or parapet several inches high. The finished structure has the form of a shallow bowl, up to three or four feet (90 to 120 centimeters) in diameter, in the midst of which rises the tower of twigs. Near some completed bowers are several rudimentary structures not unlike those of the Golden Bowerbird. Along a ridge of Mount Missim in Papua New Guinea, M. A. and S. G. Pruett-Jones found bowers regularly spaced at distances of 197 to 1,388 feet (60 to 423 meters) from their nearest neighbors. The mean interval between ninety-eight bowers was 600 feet (183 meters).

Prominent among Macgregor's Gardener Bowerbirds' ornaments are tassels of insect silk, attached to the ends of sticks in the tower, mostly within reach of the bird while he stands on the mossy carpet. One bower had forty-five of these tassels, which reminded Gilliard "of dangling cocoons as they swayed and spun in little eddies of air." Similar tassels are also hung on sticks and low plants at the edge of the mossy bowl and just beyond it. Small to medium-sized black or white objects, such as bits of charcoal, fungi, vegetable materials, and dung, are present in small numbers or often lacking. Little piles of seeds or dry brown or black berries sometimes lie upon the moss. One bower was decorated with dozens of fragments of lichens, mostly ashy white, less often blackish.

The crested owner of this bower spends much time perching incon-

Macgregor's Gardener Bowerbird, *Amblyornis macgregoriae,* male at his bower. Mountain forests of New Guinea.

spicuously well up in a tree within sight of it, at intervals bursting forth in a hodgepodge of the amazing, mostly unmelodious sounds that we have come to expect of bowerbirds. He makes noises like tapping on a box, plumage rustling, wings thrashing, limbs screeching as they rub together in a wind, insects buzzing, cloth or thick paper ripping or tearing. Mixed with these are bird calls, all highly ventriloquial. At intervals, the crested bird descends to his bower to tidy up, or to dance, hopping and moving his head from side to side, always facing the pillar of twigs, thrusting his bill into the moss at its base.

Macgregor's Gardener Bowerbird has a habit not recorded of other members of the family and rare, if not absent, in other frugivorous birds, of whatever family, in the humid tropics: he caches fruits in the vicinity of his bower. In the angle between a branch and a trunk, in the space between a trunk and a clasping vine, in a small cavity in a branch, on top of a stump, in the crown of a tree fern, and in similar situations clustered around the bower and rarely as much as fifteen yards distant, he deposits fruits from the surrounding forest. A single male might have up to fifty-five cache sites, most of which could retain only a single fruit, but one site held an accumulation of thirteen.

Since Macgregor's Gardener sometimes hangs his tassels of insect silk somewhat beyond the edge of his bower, we might suppose that he deposits fruits amid neighboring vegetation simply as an extension of bower ornamentation, which often includes berries. However, he tends to choose different kinds of fruits for caching and for ornamentation. He prefers berries of certain colors to decorate his bower, but appears not to care about the colors of his stored fruits. While displaying, he holds one of the former in his bill, but he does not use the latter. Finally, he eats his cached fruits but apparently not his display fruits, which bowerbirds commonly cast upon rubbish heaps when they decay. The Pruett-Joneses, who studied fruit-caching by Macgregor's Gardener Bowerbirds, believed that this habit enabled the male bird to remain more continuously in sight of his bower, ready to receive any female who might arrive, and guarding it from destruction by marauding males of his own kind, a not infrequent occurrence. These caches of relatively small quantities of perishable food must serve a purpose quite different from that of the larger, more durable stores that northern birds prepare to help tide them through a severe winter.

Slightly smaller than Macgregor's Gardener, the nine-inch (23-centimeter) Striped Gardener Bowerbird (*Amblyornis subalaris*) is dark brown above, more olive and ochraceous below, lightly striped on throat and breast. The male's splendid erectile crest of golden orange, brown-tipped feathers, contrasting strongly with his otherwise plain attire, is shorter than that of Macgregor's Gardener, but his bower is more elaborate and more tastefully adorned. On a wooded slope near the summit of a ridge in the lower mountains of eastern Papua New Guinea, a male Striped Gardener chooses a sapling to support his bower. Around the base of this upright he piles twigs packed with moss to make a central cone. Then he brings many more sticks to form a dome-shaped pavilion or hut with a wide opening at the front. The hollow dome may be two feet high by three feet in diameter (60 by 90 centimeters). Beneath it, on a floor composed of blackish fibers from the trunks of tree ferns, he arranges bright yellow flowers, many scarlet and bright blue berries, yellowish green leaves, and mauve-colored beetles. In front of the pavilion, on a yard enclosed by sticks, he strews brilliant scarlet fruits and sometimes a few flowers. These ornaments are frequently arranged in a definite order, yellow flowers on one side, blue berries on the other. As the fruits decay, he removes them. Little appears to have been recorded of the behavior of this wonderful bird.

Admirable as the Striped Gardener's bower is, it is surpassed in elegance by that of the most severely plain member of the family, the Brown, or Vogelkop, Gardener (*Amblyornis inornatus*), of the mountains of the Vogelkop Peninsula at the western end of New Guinea. The sexes of this ten-inch (25-centimeter) unadorned olive-brown bird are alike. Could any naturalist make a more delightful, astonishing discovery than was granted to Odoardo Beccari in the dark undergrowth of the forest in the Arfak Mountains in September 1872? It might have taken him a while to convince himself that the pavilion and garden upon which he gazed in amazement, a fit abode for Oberon and Titania, were not the work of delicately organized beings with an exquisite taste, unknown to science, but of a bird of quite undistinguished aspect. Beccari not only made a colored drawing of his discovery but described it so precisely that we can do no better than to copy his account:

> The Amblyornis selects a flat even place around the trunk of a small tree that is as thick and as high as a walking-stick of middle size. It begins by constructing at the base of the tree a kind of cone, chiefly of moss, the size of a man's hand. The trunk of the tree becomes the central pillar; another whole building is supported by it. On top of the central pillar twigs are then methodically placed in a radiating manner, resting on the ground, leaving an aperture for the entrance. Thus is obtained a conical and very regular hut. When the work is complete many other branches are placed transversely in various ways, to make the whole quite firm and impermeable. The whole is nearly three feet in diameter. All the

stems used by the Amblyornis are the thin stems of an orchid (*Dendrobium*), an epiphyte forming large tufts on the mossy branches of great trees, easily bent like straw, and generally about twenty inches long. The stalks had the leaves, which are small and straight, still fresh and living on them — which leads me to conclude that this plant was selected by the bird to prevent rotting and mould in the building, since it keeps alive for a long time . . . Before the cottage there is a meadow of moss. This is brought to the spot and kept free from grass, stones, or anything which would offend the eye. On this green, flowers and fruits of pretty colour are placed so as to form an elegant little garden. The greater part of the decoration is collected round the entrance to the nest.

Gilliard, from whose book this quotation is taken, adds that Beccari found some small applelike fruits, some rosy fruits, some rose-coloured flowers, some fungi, and some "mottled insects" on the turf.

Years later, S. Dillon Ripley found other bowers, the largest of which measured 8 feet (2.4 meters) in length and 6 feet (1.8 meters) from front to back. Built around two separate saplings, its domes rose 4.5 feet (1.4 meters) above the ground. All were extremely neat and well made. In front of each, the garden, adorned with flowers and small fruits segregated as to color, resembled a carefully tended lawn. The Brown Gardener has definite preferences and will not accept every color. Ripley dropped upon a garden a pinkish begonia, small yellow flowers, and a pretty red orchid, then watched from concealment the results of his experiment. Upon his return, the bird promptly threw aside the yellow flowers. "After some hesitation and a good many nods and looks and flirts of the tail," the begonia was also cast away. Perplexed by the red orchid blossom, the gardener took it from one of his piles of fruits or flowers to another, trying to find one that it matched. Finally, with many flourishes, he laid it on top of some pink flowers. The two colors contrasted, but this was the best match that he could find. In addition to flowers, fruits, and fungi, Brown Gardeners decorate their bowers with pieces of charcoal and black pebbles.

Brown Gardeners are known to occur only on five ranges, remote and difficult to reach, in Indonesian New Guinea. The birds on different mountains build bowers that differ greatly in construction and decorations. Early descriptions were of structures on the Arfak and Tamrau mountains. Recently, Jared Diamond found on the Kumawa Mountains bowers quite different from those previously known. Here the structures were spaced at intervals of several hundred yards on ridge crests with an eastern exposure, where they received most light in the early morning when the birds are most active. A bower contained from one to five saplings, each the center and support of a tower six to ten feet (2–3 meters) high, composed of interlaced sticks from eight to thirty-six inches (20–90 centimeters) long. These sticks were fastened together by a white, sticky substance, possibly saliva. No hut is mentioned in the published description; these bowers appear to be of the "maypole" type. Surrounding the sapling(s) was an almost perfectly circular mat of black dead moss, six inches (15 centimeters) thick and up to seven feet (2.1 meters) in diameter. A cone of the same black moss rose around the base of each sapling.

The decorations of these bowers revealed a somber taste. They consisted of black beetle elytra, hundreds of dark brown acorns, brown or gray snail shells, dark brown stones, and black twigs from small trees or tree ferns, all segregated in neat piles. As though these items were not already dark enough, the birds had painted them glossy black with some oily substance. They had also dragged into their bowers pandanus leaves up to six feet (2 meters) long and half the bird's weight, propping them against a supporting pillar or laying them flat. These gardeners also demonstrated their preference for dark objects by disdaining plastic disks (poker chips) of various bright colors that Diamond placed nearby, and by removing those that he laid upon the mossy mat. In the Wandamen Mountains, where the bowers and their decorations closely resembled those described by Beccari long ago, the Brown Gardeners showed their preference for brightly colored decorations by eagerly accepting the disks offered to them, with a decided predilection for blue, followed by purple and orange, and a total neglect of white, although in other regions white decorations are sought.

Revealing a habit not unknown among other species of bowerbirds, these gardeners stole from neighbors' bowers chips that they carried to their own. In an owner's absence, immature birds repeatedly raided his bower, pulling out sticks and tearing up the mossy mat. One adult hopped upon the shoes of one of Diamond's assistants and tried to pull off a brown shoelace and a blue sock.

In addition to regional diversity in bower construction and decoration, individual differences were frequent. This raises the interesting question of whether bower style is genetically determined or culturally transmitted, like avian song dialects and human arts and customs. In support of the second alternative, it is known that, during a long adolescence, bowerbirds of several species spend much time watching adults at their bowers, that their earliest constructions are rudimentary, and that only by much practice do they become proficient in building and decorating their bowers. They may learn by observation what bower style is most attractive to females in their region, but to confirm this more study is needed.

While Thomas Gilliard watched a bower of the more familiar type in the Tamrau Mountains, a female Brown Gardener approached. Alarmed and suspicious of the blind in which the ornithologist was hidden, she clung to upright stems beside it and peered through the window at him — a most unusual display of avian curiosity. Extremely active and suspicious, she seemed more interested in Gilliard than in the owner of the bower, who appeared to be oblivious of all outside sounds while he made frantic efforts to entice her to enter. He crouched, almost lying flat, in the dim interior of his pavilion, then scurried around inside with his tail toward the central column and appeared to look out between the bases of the stems in the wall. He "talked" in a most extraordinary manner, seeming to communicate with the female in a complex avian language. Among his utterances were catlike *meows*, explosive sounds like the drumming of grouse, rapping, ticking noises, and windy creakings in great variety. He also imitated other birds, filling the usually silent forest with their voices. The aborigines, familiar with the Brown Gardener's exceptional skill as a mimic, call him *buruk guria* (master bird). When the suspicious female departed without entering the bower, its owner continued for about a minute to cry. Then, as though piqued, he started to yank at its materials, pulling some heavy sticks from their places — rather humanlike behavior. The nest and eggs of this outstanding bird seem to be unknown.

Contemplation of the plain Brown Gardener and his exquisite bower suggested to Gilliard the inverse relationship between the development of bowers and the nuptial adornments of their builders. His "transferral effect" postulates that the more attractive the bower, the less the male bird's need of bright feathers to win females. In the course of evolution, adornments appear to have been shifted from the bird himself to the structure that he creates. Natural selection would favor this transfer, for the more cryptically colored a bird that spends much time near the ground, the less likely it is to attract predators.

Among the major enemies of the more ornate birds of paradise and bowerbirds have been the aborigines of New Guinea, who covet their plumage to adorn themselves and who, at the instigation of traders, have killed enormous numbers for export. Macgregor's Gardener Bowerbird has been heavily penalized for wearing a large golden crest that aborigines used in their headdress. The plain Brown Gardener is apparently spared this persecution. Whether humans have hunted bowerbirds long enough to affect their evolution, I do not know.

When we pass in review the bowerbirds that we have surveyed, the transferral effect is evident. Among the avenue-builders, the plainest are the Fawn-breasted and the Yellow-breasted bowerbirds, which have neither the shining, iridescent plumage of the Satin nor the rose-lilac nuchal crest of the Spotted and the Great Gray. The four-walled bower of the Yellow-breasted is the most elaborate construction of the avenue-builders. The green, blue, or red berries which adorn the bowers of this and the Fawn-breasted may be more attractive to females than the miscellaneous oddments accumulated by other avenue-builders. The three regent bowerbirds of the genus *Sericulus,* brilliantly attired in yellow or orange and black, build small bowers apparently sparingly ornamented. In the gardener bowerbirds of the genus *Amblyornis,* we have traced a series from Macgregor's Gardener, with a very large golden crest and a not especially impressive bower, through the Striped Gardener with a shorter crest and more elegant bower, to the crestless Brown Gardener, with the most charming of all bowers. Hardly compatible with the transferral effect is the Golden Bowerbird, with a very large, tastefully ornamented bower as well as handsome plumage.

The amazing variety of mostly harsh and jarring sounds that we have noticed in most bowerbirds seems strangely inconsistent with their indications of aesthetic sensibility. Should we not expect creatures who build attractive bowers and adorn them with flowers, colorful fruits, and shining baubles to prefer melodious songs and avoid uncouth sounds? Certainly, voices as flexible as theirs could produce the most dulcet phrases, copied from neighboring singers if not originated by themselves. Perhaps, in the long hours while they await a female's visit, they amuse themselves by reproducing all the diverse sounds they hear — or can invent.

Mimicry, widespread among bowerbirds, appears to be an indication of intelligence; it reveals an alert interest in phenomena that do not affect their survival. Another suggestion that bowerbirds are exceptionally intelligent birds is the way they peer into blinds, which Gilliard noticed at the bowers of several species that he watched while concealed. Although I have spent countless hours sitting in blinds before the nests of a large number of tropical American birds, some of whom accepted these structures readily whereas others were distrustful of them, none has ever, at least while I was seated within, obviously looked inside to learn what these strange objects might contain. None exhibited such intelligent curiosity.

When an individual in female plumage visits a bower, the owner often emits a volley of harsh notes and flings his display things around as though he were angry. That he is indeed in an aggressive mood appears more probable when we compare his situation with that of a monogamous territorial male, such as a Song Sparrow (*Melospiza melodia*) or a European Robin (*Erithacus rubecula*), when a potential mate first enters his territory. Since in these species the sexes are alike in plumage, he is not sure whether the newcomer is a female in search of a mate or another male encroaching upon his domain. Aggressively, he approaches the stranger. Another male will either flee or fight. A female, on the contrary, will passively endure his threats, by her behavior revealing her sex and possibly winning acceptance as his mate. Similarly, the bowerbird may be uncertain of the visitor's sex, for until several years old males resemble females. His bower is his most precious possession; he attends it devotedly, keeping it in good repair, often flying far for items to adorn it; he guards it from intruders who might carry off his treasures or harm it; understandably, he is wary of visitors. So he begins his courtship blusteringly, while the female, prudently keeping the bower walls between him and herself, waits passively until he calms down, which may take many minutes. If she waits long enough, he may mate with her.

For the question whether the adornments of male birds evolved primarily to confer advantage in contests with other males or to attract females, bowers are of first importance. Since a bower's owner cannot pick it up and wave it, like a banner, in the face of a rival, it can hardly influence the outcome of confrontations by males, but it can impress visiting females. I have found no reference to bowerbirds fighting or engaging in ritualized contests, such as are frequent among certain birds at courtship assemblies. Their rivalry takes a very different form: they wreck their neighbors' bowers, as has been reported of Satin Bowerbirds and Macgregor's Gardener Bowerbirds and possibly occurs in other species. This behavior sets bowerbirds sharply apart from birds who court in assemblies and cooperate to facilitate the females' choice of nuptial partners. Bower destruction, together with the usual wide separation of bowers, makes it more difficult for a female to compare potential partners and choose between them. However, bowers of the Satin bird, and probably others with the exception of the more elaborate of them, are readily reconstructed.

Like other frugivorous birds, bowerbirds must wander widely as, now here, now there, a tree or shrub ripens its fruit. Despite rather wide separation and occasional wrecking, bowers of the more abundant species appear not to be too far apart for a female to become familiar with several of them and to choose between them, as in the exploded, or dispersed, leks of certain birds of paradise and cotingas. If females enjoyed little opportunity to choose, sexual selection would be so impeded that bowerbirds might not have evolved the handsome plumage of some of them, or transferred their adornments from their bodies to their bowers, as others have done.

References

Borgia 1985a; 1985b; Diamond 1986; Gilliard 1956; 1959b; 1959c; 1969; Loffredo and Borgia 1986; Marshall 1954; Pruett-Jones and Pruett-Jones 1985; Ripley 1942; Vellenga 1970; 1980; Warham 1957; 1962.

12.
The Courtship of Lyrebirds and Dancing Whydahs

After reminding the reader that the iridescent green streamers of the Resplendent Quetzal (*Pharomachrus mocinno*), like the two hundred feathers of the gorgeous fan of the Blue Peacock (*Pavo cristatus*), do not belong to the tail but are upper tail coverts, I shall venture to assert that the male Superb Lyrebird (*Menura novaehollandiae*) wears the most elegant of all tails. Certainly it is one of the avian tails that have diverged most widely from the tail's original function of stabilizing flight, and perhaps none has been more frequently pictured. The bearer of this wonderful tail is a pheasant-sized bird, up to thirty-nine inches (100 centimeters) long, plainly attired in brown above and grayish brown below. His legs are long and stout for running and scratching in the ground. His only adornment is his tail of sixteen feathers. The two outermost, shaped somewhat like a Grecian lyre, are S-shaped, with one vane greatly reduced. The broad inner vane has alternating dark and light transverse bars that make it appear deeply notched. Next within is a single long, thin plume on either side. The twelve interior feathers, known as filamentaries, have long, loose barbs, devoid of barbules, that impart a lacy aspect. The smaller female has a shorter, simpler tail.

Superb Lyrebirds are found only in the mountains and foothills of southeastern Australia, from southern Victoria to southeastern Queensland. They forage beneath forests of eucalypts and southern beeches (*Nothofagus*), scratching and digging in the ground or tearing open rotting logs for small invertebrates such as worms, crustaceans, centipedes, spiders, and insect larvae. They are poor flyers, able to cover long distances only by gliding downward, or to flit clumsily between the branches of trees. Nevertheless, they roost high in the crowns, which they reach by jumping from branch to branch.

Superb Lyrebird, *Menura novaehollandiae*, male displaying.
Southeastern Australia.

Male lyrebirds mature slowly. In their early years, small groups of them wander and display together, not without frequent mutual aggression. At the age of seven years, when at last their tails are fully developed, they claim marginally overlapping territories of about 6 to 8.5 acres (2.5 to 3.5 hectares) which, especially in winter, they defend by displaying, chasing intruders, and singing. In each territory they scratch up many mounds of earth from three to five feet (1 to 1.5 meters) in diameter. A single male may have up to twenty of these hillocks, upon which he displays and mates. Within his domain as many as four, and possibly up to seven, females establish smaller territories, which they defend against other females but permit the male to wander unchallenged through them, to forage or visit the mounds where he sings and displays. The males' mounds are usually on ridges or where fallen trees or dying limbs permit more light to reach the ground. The females' territories are more often in gullies, where food and nest sites are more abundant.

At the height of the breeding season, a male Superb Lyrebird may devote half the daylight hours to singing, in trees, on rocks or fallen logs, or on the ground. The song of his own species is so loud that on days with little wind it is audible over half a mile (1 kilometer) away. This song has developed many distinctive local dialects. The singing lyrebird avoids monotony in his long-continued outpourings by accurately imitating the songs and sounds of other birds. Since lyrebirds breed in midwinter, when most neighboring birds are sexually inactive and sing little, their renditions of these songs probably cause little confusion. A local population of lyrebirds may mimic the notes of up to sixteen other species of birds, plus the vocal and wing sounds of a flock of flying parrots, the barking of dogs, and, reputedly, the horns of motorcars.

The site of a Superb Lyrebird's displays is one of his mounds. Standing there, he swings his expanded tail upward and forward to form a shimmering, silvery white canopy, framed by the curiously shaped, barred outer feathers, above his plain brown back and extending forward well beyond his head. At the same time, he turns slowly and emits a continuous stream of loud, melodious notes. His display to a visiting female is somewhat different. Now, instead of fanning out his tail held horizontally over his back, he contracts it until the lyrate outer plumes are more or less parallel and the lacy filamentaries are bunched, rapidly quivering, between them. He accompanies this display with a continuous, clicking *cric-cric-cric*. As the performance reaches its climax, he traces semicircles around the female with such short, rapid steps that he seems to glide. Then he leaps back and forth in time with a rhythmical call that ends with two far-carrying, bell-like notes.

Female lyrebirds receive no help during the several months that they take to build their roofed nests with a side entrance on earthen banks, rocky ledges, boulders, or in clumps of grasses or sedges, usually within six feet (2 meters) of the ground but occasionally up to seventy-two feet (22 meters) in trees. The base and sides of the bulky structures are made of sticks compacted with moss and strips of bark; the chamber is lined with fine rootlets, with sometimes a little green moss on the roof. The average weight of these structures is thirty-one pounds (14 kilograms). Each nest receives a single gray egg, marked with dark brown, which the hen incubates for about fifty days, an amazingly long period for a passerine bird. The egg takes so long to hatch because it is incubated for somewhat less than half of the daytime. While the hen is absent, foraging, for three to six hours on cool winter mornings, the embryo becomes so thoroughly chilled that its development is interrupted.

The chick, hatched with thin black down on its upperparts, is fed only by its mother during the forty-seven days that it remains in its roofed nest. After its departure, it accompanies its mother and receives part of its food from her for up to eight months. Hardly less unusual than the male lyrebird's tail and displays is the female's nesting — the ponderous covered nest, midwinter breeding, long incubation and nestling period, prolonged maternal care of the fledged young. The two facets of the reproductive program are related: if, instead of displaying his elegant tail, the male lyrebird helped to rear the young, they might develop much more rapidly. In summer, after the breeding season, the lyrebirds molt and then gather in flocks of up to fifteen birds, including four adult males.

The only other species in the lyrebird family, the Menuridae, is Albert's Lyrebird (*Menura alberti*), which has a smaller range in northeastern New South Wales and southern Queensland, has a tail slightly less ornate than that of its congener, and has been less thoroughly studied.

A very different bird with a very different tail is Jackson's Dancing Whydah (*Euplectes jacksoni*), confined to the highlands of Kenya, and the only member of the weaver family known to display at a court. The black male is sparrow-sized, with a full tail of vertically expanded, arching plumes longer than his body. On the open veldt, he selects a thick tussock of tall grass, around which he makes a circular runway by clipping off short lengths of surrounding grass with his bill. Before his court is completed, he begins to display on it, at the same time nipping off more bits of grass and dropping them, until his track becomes four feet (1.2 meters) in diameter, covered with the dried and shriveled grass blades that he does not remove but tramples down. In the sides of the central tussock he opens two stalls or recesses.

When a male whydah comes to his court, he walks around the track, clipping off a bit of grass here and there. Then, crouching, he shakes his body from side to side and emits a sizzling sound. Next, he pushes into one of the niches in the central tuft. He raises his head- and neck-ruffles, throws his head well back, and swings his tail forward until it almost touches his head. Swaying from side to side, he moves jerkily in and out of the recess, to the accompaniment of wheezes and rattles. After this, he starts to dance or jump, rising two or more feet into the air with quivering wings and tail cocked up, with the exception of two feathers that remain horizontal. After five or six upward leaps, he may rest or repeat the ground display at the tuft. All his movements are vigorous.

Jackson's Dancing Whydah, *Euplectes jacksoni,* male displaying.
Highlands of Kenya, East Africa.

He displays from early morning until about an hour before noon, then again, more energetically, between about half-past four and five o'clock in the evening.

The courts of Jackson's Dancing Whydah are grouped in an assembly. One year, V. G. L. Van Someren found three courts, two fairly close together, the third a little way off. Ten sparrowlike, cryptically colored, short-tailed females appeared to be associated with the three males, and seven nests were discovered in the vicinity. Each was built beside a grass tuft, in a little hollow made by cutting and pulling away old, decaying grass. The female lined the depression with dry grass blades, then pulled down living grasses and, weaving them into the dry materials, roofed her structure, which became a ball with a side entrance, camouflaged by the green blades that covered it. In it she laid three pale blue eggs, spotted with brown, which hatched after twelve or thirteen days of incubation by the female alone. In four days of watching at each of two nests,

Van Someren never saw a male approach. Fed chiefly with regurgitated grass seeds gathered in a neighboring swamp, the young whydahs left their nests when they were about seventeen days old and could hardly fly.

With several females nesting in his defended territory, the Superb Lyrebird is polygynous, with a penchant toward promiscuity. With several males making their courts and displaying fairly close together, and no defended territories in which females nest, Jackson's Dancing Whydah appears to court in an assembly, like Ruffs and manakins. Despite the differences in their mating systems, both the lyrebird and the whydah have developed ornamental tails. The great differences in their tails might be due to different preferences of the females, but more probably they are expressions of deep-seated hereditary tendencies in their respective families. Both the lyrebird and the whydah conform to the reproductive pattern that most often produces lavishly ornamented male birds who never attend nests. Whether they practice polygyny or court in assemblies, males outstandingly successful in attracting females transmit their genes to more descendants than most monogamous males can beget, and this leads to the relatively rapid evolution of whatever features give them this advantage. A relative of Jackson's Dancing Whydah, the polygynous Long-tailed Whydah (*Euplectes progne*) is black with red-and-white epaulets and a tail twenty inches (50 centimeters) long. He flies slowly and low over the territory where from six to twelve of his females nest amid tall grass, in roofed structures with a side entrance.

References

Lill 1979; 1985; Rowley 1975; Van Someren 1956.

13.
The Courtship of Ducks

Many male ducks are beautiful. The widespread Mallard (*Anas platy-rhynchos*), the Green-winged Teal (*A. crecca*), and the Falcated Teal (*A. falcata*), to name only a few, are handsome birds; the Wood Duck (*Aix sponsa*) of North America is even more elegant; its oriental counterpart, the Mandarin Duck (*A. galericulata*), seems overly ornate. The females of the more colorful drakes are nearly always much plainer, and rear their families without male assistance. In all these ways, ducks, especially the migratory northern species, differ from the great majority of aquatic birds of both fresh water and salt, in which the sexes are alike and both participate in nesting. What causes these differences? What factors have contributed to the beauty of drakes? A study of their courtship might help to answer these questions.

Drakes of migratory species that nest in middle or high northern latitudes begin to court while they forage and rest in flocks on waters where they winter farther south. Near the Pacific coast of Mexico at the end of December, Logan J. Bennett watched drakes of the Blue-winged Teal (*Anas discors*), just beginning to acquire their nuptial plumage, mildly bowing and dipping before unpaired females. As the teals migrate northward with the spring, they court more ardently on the ponds where they rest before continuing onward.

Wood Ducks form pairs while they winter in the southern United States. Their early mating is favored by early acquisition of their nuptial elegance. Like many other ducks, male Wood Ducks desert their mates during the incubation period, leaving the latter to care alone for the ducklings they have sired. The drakes retire to some sequestered pond or lagoon where they remain while, becoming flightless, they molt into the eclipse plumage. They wear this plain attire only briefly before they

Wood Duck, *Aix sponsa,* male. Eastern and central North America from
southern Canada to Florida and Texas; far western North America from
southern British Columbia to southern California.

molt again, regaining the colors in which they will breed many months
later. This molt is often completed by September or October, before the
drakes disperse from the ponds where they pass their flightless fortnight,
soon to undertake their southward migration. Thus, they are splendidly
prepared to court the females in midwinter, far from the regions where
they will breed. After a pair is formed, the drake closely follows his partner wherever she leads, to the foraging site, to still water where they
pass the night, to the spot where they loaf by day.

In the spring, when the ducks begin to move northward, the drake
continues faithfully to follow his mate. She usually returns to the locality
where she nested the preceding year, perhaps to the same hollow tree or
nest box, or, if a young duck, to near her birthplace. His attachment to
her may lead him to a spot far from his own birthplace. When a drake
reared in Louisiana pairs with a female from Michigan, he follows her
to the northern state. Conversely, a drake from Michigan may join a
duck hatched in Louisiana and remain there with her through the early
weeks of the nesting season. After he has watched his partner choose a
nest cavity, lay her eggs, and start to incubate, the pair bond dissolves.
In the following year he may win a female of different origin, with the
result that he may in successive seasons breed in regions far apart, perhaps one year in Illinois and the next in Maine. The male Wood Duck's
attachment to a mate, of whatever geographic origin, rather than to a
locality, creates the situation known as panmixia, with interesting ge-

Mandarin Duck, *Aix galericulata,* male. Eastern Asia and Japan.

netic results. A male who by mutation gains an advantage in courtship, perhaps intensification of his colors, may spread his genes over a great area. Probably this situation has helped male Wood Ducks to become so outstandingly beautiful. The behavior of Wood Ducks is not unique. While wintering or migrating, drakes of a number of other species form alliances with females who, returning to their own birthplace, lead their male partners to regions new to them. Female ducks are more philopatric than males.

The courtship displays of ducks appear to be less diverse than those of manakins, cotingas, and birds of paradise. Certain movements are repeated in the demonstrations of various species. Head waving or head bobbing is widespread. Frequently the head is thrown rearward until it rests upon the drake's back. His neck and head may be stretched forward over the water toward the female whom he courts. Or he may extend his neck upward, with bill pointing toward the sky. His neck is often swollen with air. Dipping the bill into the water is a feature of many displays;

splashing water is frequent. The drake may rear upright in the water to show his underparts. He may pursue the female in the air or on the water. He may perform silently or make loud, mostly unmelodious sounds, variously characterized as grunts, groans, quacks, coos, and subdued whistles. Some of these gestures are repeated, usually with less intensity, by a responsive female. Often two or more drakes simultaneously court a female. She shows definite preferences and may press close to the drake of her choice, attack one who fails to please her, or incite her partner to do so. The drake who wins a bride leads her away from the throng of ducks who often surround the courting pair, interrupting the proceedings before they can be driven aside. In a more secluded spot, the male mounts the female, who is often wholly submerged by his weight during the fraction of a minute that coition requires.

After the foregoing generalizations, let us look at a few specific examples of courtship. Bennett gave a good account of the Blue-winged Teal's displays. After beginning, as we have seen, to court while still in their winter home, these ducks continue their demonstrations during their northward migration, often courting in the air, coursing and dipping over the waterways between their main flights. However, the most interesting phase of their courtship occurs on the water and surrounding shores or mud flats. By creeping through the grasses up to the edge of an Iowa pothole, Bennett was often able to watch the ducks' nuptial activities. A male and female, or several of each sex, swam slowly about, bowing to each other, often in unison. Sometimes either a male or a female bowed singly, apparently ignored by the others. For intervals of several minutes, a bird might bow once every two seconds, its head and neck moving up and down and slightly forward. For hours together, with brief intermissions for eating and resting, the teals swam around each other and bowed. Birds of both sexes often courted peaceably together for hours before one duck chased another from the female. After several days of persistent chasing, one of the drakes finally won a partner. Often males in flight uttered peeplike whistles, but they were never heard to quack. These Blue-winged Teals courted chiefly from dawn to ten o'clock in the morning, and from four in the afternoon until nightfall. Similarly, Jessop B. Low found Redheads (*Aythya americana*) courting most actively for 2 to 2.5 hours after dawn, and in the evening for 1.5 to 2 hours before nightfall. Redheads pursuing females flying above the marshes try to force them to alight on the water in order to display to them.

John G. Millais told how Common Pochards (*Aythya ferina*) court. In March, flocks of these ducks arrive in Great Britain and settle on large expanses of open water. On the first warm day, drakes start to court, four or five of them crowding around one female, who in turn circles around one of them, dipping her bill in the water, stretching her neck low on it, occasionally voicing a hard *kurr-kurr-kurr*. The males continue to utter a curious groan, mingled with a soft, low whistle, audible for only a short distance. They toss head and neck rearward until the hindhead touches a point between the shoulders, repeating this constantly in the earlier phase of courtship. More often they distend the neck with air, raise the head, and emit a groan as the air is released. Display reaches its climax when a male, lying very flat on the water and stretching his head and inflated neck far forward, approaches a female. In this attitude he frequently turns his head sideways, the better to show its beauty. Two or three drakes may lie together before a female. In intense excitement, the iris of a male pochard's eye becomes a blaze of rich lacquer red, obscuring the pupil.

Anthony J. Erskine described and illustrated the courtship displays of the little Bufflehead (*Bucephala albeola*) in British Columbia. A drake courting a female in the presence of another male flies a short distance over or past her, his head and fluttering wings held below his oddly humped body. Just before landing, he rises nearly upright and advances his feet so that, like water skis, they reduce his velocity. In this attitude, the striking black-and-white plumage of his back and his pink feet are prominent. As he settles into the water, he bows forward and flicks his wings above his back. He turns toward the female and approaches her, rapidly bobbing his head with mechanical jerks. His performance draws other drakes, sometimes as many as eleven, who fly or swim in and alternately bow their heads toward the female and chase each other, splashing water widely. Gradually, drakes drop out of this confused melee, leaving only one or two, one of whom swims rapidly away, followed by the female uttering a guttural *ik-ik-ik-ik*. After pairing, Buffleheads appear to recognize their mates, but the drakes frequently leave their partners to court other females, including those already paired — a habit widespread among ducks. Buffleheads do not breed until they are nearly two years old.

Despite the very different settings, one who reads the foregoing accounts is reminded of singing assemblies of hummingbirds in tropical woodland, courtship gatherings of manakins, or perhaps leks of Ruffs on northern meadows. The feature common to all these diverse assemblies is the simultaneous competition by a number of males for females whose choices they cannot compel. To be sure, a flock of ducks engaged in aquatic courtship is less orderly than a courtship assembly on dry land. An expanse of open water does not favor the establishment of closely adjacent, clearly demarcated territories such as we find in terrestrial courtship assemblies. Moreover, ducks that migrate far would hardly have time to claim territories, and become familiar with them, before they must pair for early nesting. Hence the disorder, the crowding around and interfering with each other's courtship displays, to a degree that we rarely find in the courtship assemblies of terrestrial birds. Although discussions of leks and other courtship assemblies commonly fail to mention ducks, I include their spring gatherings in this category because doing so helps us to understand why drakes of many species are so handsome.

In these gatherings, intersexual selection is active, along with a certain amount of intrasexual selection as males compete with each other. Since males of northern ducks do not normally incubate or attend the ducklings, they have little need of cryptic plumage, such as females wear, but are free to become colorful as a result of female choice. Because northern ducks commonly lay from eight to twelve eggs and sometimes more, a male who mates with a single female has the reproductive potential of a hummingbird, manakin, or bird of paradise who wins half a dozen females who lay only one or two eggs. The large clutches of ducks should favor the rapid multiplication of the genes of a highly successful drake.

Ducks confined to the Southern Hemisphere differ in important ways from those of the Northern. John Warham reported that the sexes of the attractive Pink-eared Duck (*Malacorhynchus membranaceus*) are alike in plumage. Remaining with his mate throughout the breeding season, the male escorts her to the nest in a hole or crotch in a tree and possibly shares incubation. He helps his mate to call newly hatched ducklings from the nest to the water, and accompanies them and their mother as the family swims away.

During a year with the ducks of Argentina, Milton W. Weller noticed a tendency toward permanent pairing, with drakes accompanying broods of ducklings. Sexual differences in plumage were less frequent and pronounced than in northern ducks, and males did not go into eclipse after the breeding season. Their courtship appeared to be less intense but more prolonged than that of northern ducks. Frank McKinney listed nine species of dabbling ducks (*Anas*) of the Southern Hemisphere in which drakes usually or probably accompany ducklings, as rarely occurs among northern ducks. Similar differences between tropical and northern species are found in other avian families, and appear to be closely associated with the prevalence of permanent residence or reduced migration among tropical and subtropical birds.

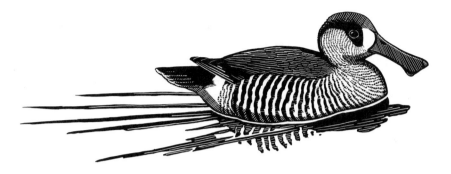

Pink-eared Duck, *Malacorhynchus membranaceus*. Sexes similar.
Most of Australia, Tasmania.

Ducks, and other birds, do not invariably choose the most handsome male available to them; other factors may influence their choices. Frank Finn, an early student of sexual selection, watched semidomesticated Mallards in Regent's Park, London, in the pairing season. The drakes varied greatly in color; on some the rich chocolate breast was lacking; others had even a slate-colored head instead of the usual brilliant green. Yet these "off-coloured" drakes succeeded in winning and keeping mates, while correctly dressed males remained bachelors. One gray-breasted drake was even able to indulge in bigamy. The successful drakes were able to drive away their unmated rivals, "a proceeding in which their wives most thoroughly sympathised, as their gestures plainly showed." Apparently, the superior size or vigor of the plainly attired drakes compensated for their dull plumage.

Free choice of mates is extremely important to ducks, as Cynthia K. Bluhm demonstrated with captive Canvasbacks (*Aythya valisineria*) raised from eggs collected in the field in Manitoba and Saskatchewan. After giving the birds an opportunity to court and pair in a flock containing equal numbers of males and females, she segregated them, one male and one female together, in isolated cubicles containing everything they needed. The control group contained nineteen firm pairs spontaneously formed. Ten females were separated from their mates and given drakes taken from different pairs. Another group of twelve pairs was formed by placing females who had been courted, but had not yet chosen partners, with males arbitrarily assigned to them. The results were spectacular. A few hours after they were placed in the compartments, the freely formed pairs fed and courted together, and soon nearly all the females laid eggs. Pairs formed by joining a female who had been courted, but had not yet chosen a mate, coexisted more or less amicably but never produced an egg. Females separated from their chosen partners and confined with other drakes pecked and chased them. Although larger and heavier than their persecutors, the drakes did not retaliate; most succumbed to this harsh treatment. When, after an interval with drakes assigned to them, females were reunited with the partners they had earlier chosen, they accepted them and laid eggs. Most of these captive Canvasbacks retained their mates for only one season; but 14 percent of the pairs remained intact for two years; one, for seven years.

It has long been known that enduring pairs are formed by geese, swans, and ducks of which the sexes cooperate in rearing their young. Pairs or families of swans and geese migrate together for long distances, and mated adults remain together throughout the year. The drake's habit of abandoning his mate after she has laid her eggs, widespread among northern ducks, appears to terminate the pair bond. However, this does not always follow. Although drakes of Barrow's Goldeneye (*Bucephala islandica*) desert their incubating mates, molt, and migrate to the wintering area while the females are raising their families, the partners may later recognize one another and reunite. Jean-Pierre Savard told how pairs of goldeneyes, who had been separated for four months, renewed their bonds after the females joined the males on the coastal waters of southern British Columbia. After passing the winter together, these reunited couples returned together to their nesting ground two hundred miles (320 kilometers) away. Strong attachment to a chosen wintering area as well as to a breeding area helps to preserve pair bonds from year to year. Since such fidelity to known localities for both nesting and wintering is widespread in ducks (as in other birds), enduring pair bonds may prove to be more common among waterfowl than has been thought. Although the aquatic courtship of ducks resembles terrestrial courtship in assemblies not only in the opportunity it provides for female choice but also in promoting the evolution of beautiful male birds who mate with plainer females, the mating of ducks is rarely as fleeting as that of terrestrial birds in their assemblies, and sometimes they form enduring bonds.

References

Bennett 1938; Bent 1923, 1925; Bluhm 1985; Erskine 1972; Finn 1907; Low 1945; McKinney 1985; Millais in Bent 1923; Savard 1985; Stewart 1959; Warham 1958; Weller 1968.

14.
Courtship Assemblies and Avian Splendor

The birds whose courtship we have just surveyed — especially humming-birds, manakins, cotingas, birds of paradise, bowerbirds, lyrebirds, and ducks — wear some of the most splendid plumage, or build some of the most elaborate constructions in the whole avian class. What factors have contributed to this result?

In recent years, ornithologists have patiently counted the number of matings of recognizable male birds in courtship assemblies or at bowers. Some of the results of their studies are summarized in Table 2. The first line in this table shows that twenty-six males in a lek of Sage Grouse mated with females a total of forty times. These matings were far from being equally distributed among the males who competed for them. The two most successful individuals mated, respectively, thirteen and fifteen times, together accounting for 70 percent of all matings, while fourteen males did not win a single hen. Together, the remaining ten males mated only twelve times. In species after species, we find that one or a few favored males attracted most of the female visitors, while many were neglected by the females. This probably does not mean that during a whole season their strenuous efforts to win a partner were unrequited. I doubt that all the male participants in any of the assemblies were under surveillance continuously enough to prove that. Moreover, such great disparity in the number of matings by males in assemblies does not occur in all species. At gatherings of Long-tailed Hermit Hummingbirds, Gary Stiles and Larry Wolf learned that all males, including subordinate individuals in peripheral territories, had a fair chance of winning a female.

The selective mating that this table reveals leads to the rapid diffusion through a population of the genes that contribute to an outstanding male's success. Many or most of his male offspring should inherit what-

Table 2.   *Mating Success of Males at Courtship Assemblies or Bowers*

Species	Number of Males	Total Matings	Males Failed to Mate	Matings by Top Male(s) No.	%
Sage Grouse 1	26	40	14	13, 15	70
Sage Grouse 2	30	87	21	25, 41	76
Black Grouse 1	6	24	3	17	71
Black Grouse 2	10	25	5	7, 9	64
Black Grouse 3	15	36	5	15	42
Black Grouse 4	9	26	5	9	35
Capercaillie	5	24	3	22	92
Greater Prairie-Chicken	9	30	6	21	70
Buff-breasted Sandpiper	22	22	14	7	32
Ruff 1	22	47	16	19	40
Ruff 2	15	82	6	11, 13, 14, 17, 19	90
White-bearded Manakin	12	28	9	18	64
Golden-headed Manakin 1	13	87	3	12, 14, 14, 14, 22	87
Golden-headed Manakin 2	16	78	6	13, 15, 18	59
Lesser Bird of Paradise	8	26	6	25	96
Raggiana Bird of Paradise	4	29	0	15, 4, 7, 3	100
Lawes' Six-wired Bird of Paradise	15	22	7	9	41
Satin Bowerbird	22	207	1	20, 21, 21, 25, 33	58

Sources: Payne 1984; Borgia 1985b, Beehler 1983b, 1988.

ever feature(s) made him so attractive to females, becoming in turn highly successful members of their assemblies; and this situation, repeated generation after generation, should lead to relatively rapid evolutionary change.

When we ask what chiefly contributes to the outstanding males' success, the answer is evidently lavish adornment plus skill, and persistence in displaying it, or, in the case of bowerbirds, attractive bowers, for these are the features that selective mating has promoted. When we recall the powerful attraction of supernormal stimuli, or any exaggeration of the signs to which birds commonly respond (Chapter 5), we can hardly doubt that a mutation which increased the splendor of a male's plumage, or his dexterity in displaying it, would give him an advantage over his competitors, and successive mutations in the same direction should intensify these characters.

However, the situation is not as simple as this. As a rule, conspicuous genetic mutations occur infrequently; populations of birds tend to uniformity; all the fully adult members of a courtship assembly may appear much the same to us — the outstanding exception being the Ruffs. What, then, makes certain individuals in an assembly so much more successful in winning females than their neighbors? Among the factors that confer advantage is central location in a large assembly, as in the Sage Grouse and the White-bearded Manakin. Just as, among tightly clustered colonial birds, the youngest members of a colony nest in peripheral sites and, as they grow older, work their way inward, so, in populous assemblies, the latest recruits settle about the edges and penetrate as rapidly as they can toward the center. Competition for position may be keen but nearly always takes the form of harmless confrontations by neighbors. If severe injuries were often inflicted, loss of members would dissolve leks.

Some naturalists have contended that brilliant colors and profuse plumage have evolved primarily to intimidate rivals in contests for rank in courtship assemblies rather than to attract females. It will be recalled (Chapter 10) that while Mary LeCroy and her associates watched the prolonged courtship display of a full-plumed male Goldie's Bird of Paradise, young plumeless males slipped in and repeatedly mounted the female whom he was addressing. This observation is cited in support of the view that the adornments and displays of males serve them chiefly in contests with other males for the status or position that confers advantage in winning females; their effect upon the females themselves is regarded as having only secondary importance. However, the observation is difficult to reconcile with either interpretation of the significance of male adornments. What does a mature male gain by winning high status among his peers if he permits young males to precede him in coition? And what do his splendors and blandishments bestead him if he allows unadorned youngsters to intrude? What this episode suggests is that nuptial display of splendid adornments before a fascinated female may become so intense and protracted that opportunists may reap advantage. In the absence of privacy assured by large, well-defended territories, prolonged, elaborate courtship may become self-defeating. In many courtship assemblies, mating is quickly accomplished after a female has signified her readiness. Moreover, similar intrusions by immature males are certainly not usual at courtship assemblies and may be exceptional even at that of Goldie's Bird of Paradise, which was watched for only eighteen hours.

A male bird's adornments may be not without influence upon the outcome of his confrontations with other males for status. A difficulty here is that, lacking mirrors, a bird cannot see the full splendor of his own plumage and compare it with that of his adversary, perhaps deferring to his superior attire — as a female can compare several males who vie for her attention. The psychic effects of colors and ornaments depend largely upon the context in which they are displayed. In a dangerous situation, they may intensify dread; in a friendly atmosphere, they may foster admiration or love. Men have traditionally worn their best clothes to go courting; and when warfare was frequently a direct confrontation of warriors instead of mechanized slaughter at a distance, they adorned themselves, often lavishly, for battle. The same adornments that help to impress a rival of a male bird's superiority may favorably stimulate a female in search of a partner. A bird who has been contending with a competitor cannot quickly change his clothes to address an approaching female, although he may alter his manner of displaying his elegance. As Julian Huxley pointed out, bold patterns visible at a distance predominate in threat displays, whereas nuptial displays are more likely to reveal delicate details.

In any case, the view that the primary function of the males' adornments is to confer advantage in contests with other males for status in a hierarchy does not account for the evolution of ornamented bowers, which cannot be picked up and flaunted in the faces of rivals. Moreover, a male trying to dominate another would surely not hang head-downward before him, as in the inverted displays of some of the most beautiful of the birds of paradise. If we seek a theory that can account consistently for elegant plumage, inverted displays, and tastefully decorated bowers (including the transfer of adornments from the birds to their bowers), we will prefer the classical theory which attributes these embellishments to the females' choices of nuptial partners. Although it is true that both intersexual and intrasexual selection are active in courtship assemblies, the former is mainly responsible for the splendor of their participants.

Success in winning females in courtship assemblies increases with age and the experience it brings. Recent studies by P. Shaw, Bruce H. Pugesek, Kenneth L. Diem, and others have demonstrated that Blue-eyed Cormorants (*Phalacrocorax atriceps*), California Gulls (*Larus californicus*), and several other long-lived seabirds tend to seek mates of their own age. The prevalence in these species of pairs of equal-aged birds implies the ability of birds to recognize the ages of their associates. Older pairs of monogamous birds usually nest more successfully, fledging more young, than pairs with less experience. Among birds that breed cooperatively, the oldest male is head and leader of his group, with others ranked below him in the order of their ages. Similarly, in courtship assemblies, older males are likely to occupy the preferred stations, perhaps won by years of slow advance. Females do well to choose older males to sire their nestlings, for in a state of nature, without institutions to prop up the weak and the faltering, survival itself is a measure of competence and vitality. This is especially true when the bird wears bright colors that make him more conspicuous to predators, or extravagant plumes or dangling wattles that may impede his movements. Thus by choosing older males with central stations and possibly the richest plumage and most alluring displays, females at courtship assemblies select genes that will make vigorous offspring.

One may ask why male birds continue to attend assemblies where others are winning nearly all the matings. Why do male blue-backed manakins (*Chiroxiphia* spp.) join in displays only to stand aside and watch their dance partners, or one of them, mate with all the females that they have together attracted? Would subordinate males not do better for themselves if they spread out and each, independently, used all his arts to entice females to his defended territory?

These questions are not easily answered. For the species, courtship assemblies have definite advantages: they are readily found by females with developing eggs because a number of displaying males make them conspicuous and, the habitat remaining favorable, are in the same spot year after year. At them a female can readily compare a number of competing males and choose the one who most impresses her. What, then, does the individual male gain by joining an assembly where for years his chances of winning a female may be slight? Probably he is safer in the assembly. Undoubtedly, a number of birds performing and calling together in a well-known locality are more readily found by predators than a male displaying alone; but in any case he must make himself conspicuous to attract females. Probably a number of vigilant participants in the assembly more than compensate for increased conspicuousness — just as a bird not only feels safer in a mixed foraging flock than when alone, but is actually safer because the many watchful eyes and voices quick to give warning make it more difficult for a predator to surprise any of the flock's members.

Moreover, a male bird might find it more pleasant to pass his days with others of his kind than to spend long, solitary hours waiting for a female to approach an isolated station; he might enjoy confronting rivals in ritualized encounters, or simply resting among them in intervals of inactivity. Manakins often perch close together when business is slack in their assembly. In cooperating with other males to attract females for whom individually they compete, males in courtship assemblies may find a substitute for the mate and family of which the vagaries of evolution have deprived them. And perhaps, in the long run, and with luck, the subordinate male will beget as many offspring or more than he might engender if he performed in solitude. If he survives long enough, the male neglected by females while stationed at the edge of a lek may work his way to a central position where he wins many. The subordinate manakin of a dancing pair or trio may become the dominant bird at the same station if he outlives the probably older alpha member of his team, and in turn enlist younger associates to join him in exhibitions that win females for himself. Success in the evolutionary game is measured by the number of his progeny.

In the birds that we have been considering, individuals in female plumage are commonly much more numerous than those in the full regalia of adult males. This does not necessarily indicate an unbalanced sex ratio because males may live for years in the dress of immaturity. Male Satin Bowerbirds do not acquire the iridescent violet-black plumage until they are from four to seven years old. Male Superb Lyrebirds are equally slow to develop the tails for which they are named. Male Long-tailed Manakins need three or four years to develop full adult plumage, male Blue Manakins two or three years, and male White-ruffed Manakins more than one year. Why do these small birds delay so long, after they cease to grow in size, to acquire the definitive plumage of their sex? What factors cause them to retain female colors to an age far beyond that at which other birds of about the same size, with very different mating habits, wear adult plumage?

An obvious answer is that since in courtship assemblies one male can inseminate many females, only a relatively small number of fully mature males is needed; if all males promptly entered the breeding population, their number would be excessive. However, I believe that we must look for more positive advantages of deferred adult plumage in birds with these habits. By retaining the more subdued colors of females and juvenile males, the youngsters remain less conspicuous to predators. They have time to perfect elaborate courtship displays by practice, away from the assemblies, or at the courts of adult males while the owners are absent or even present. They can associate closely with displaying adults, learning from them, without arousing the opposition that intruding males in full regalia would encounter. If they acquired adult plumage and became regular members of the assemblies at an earlier age, they would have little success in a system dominated by elders. Nevertheless, at least some of them become sexually active while still in juvenile or transitional plumage, and occasionally, when adult males are absent or unobservant, they mount soliciting females. During the long years before bowerbirds don full adult attire, they practice building bowers, a difficult art which apparently must be learned, although doubtless it has an innate foundation.

Apparently, no insuperable obstacle would prevent the more rapid maturation of manakins, birds of paradise, and others with similar mating habits if this were advantageous to themselves or their species. Evidently, the delayed maturation of these male birds is adaptive. It is instructive to compare the situation in these birds with that in penguins, albatrosses, gannets, boobies, auks, puffins, and other long-lived oceanic birds. The sexes being alike, they early acquire adult colors, but they delay breeding for periods comparable to those of the birds that have engaged our attention. This, too, appears to be an adaptive postponement of maturity.

Because females resemble young males who sometimes display in courtship assemblies, it would be hazardous to assert roundly that in many species females never display. Nevertheless, females of the Orange-collared Manakin and other species of *Manacus,* who jump with males back and forth over their bare courts, are exceptional among birds of this category. Nearly always, undoubted females are passive spectators of the displays of males whom they visit for insemination. Often they appear to be fascinated by them. To these watchful, demure, unadorned females and their choices, we owe a substantial part of the feathered world's splendor.

How do these females recognize the males of their species? Nestlings attended by both parents become familiar with their appearance and voices at an early age. If the sexes differ in plumage, a young female seeking her first mate should have no difficulty recognizing a male who resembles her father, nor a young male in accepting a female who looks like his mother. If the sexes are confusingly similar in appearance, recognition of the species should be easy but learning the sex of a prospective partner might take more time. But young birds never attended by a father may not see a male of their kind until they become independent of maternal care and are on their own, often leading a more or less solitary life. When ready to breed, how do the young females recognize as appropriate partners male birds who often look very different from the only parent they ever knew?

Although I can give no definite answer to this question, several possibilities occur to me. The young female may have an innate image of the male of her kind, or at least of his outstanding features, as appears to be true of certain ducks, raised by mothers unattended by an adult male. The young female may recognize males of her species by calls used by both sexes that she has heard from her mother. For example, the courting male Blue-crowned Manakin utters both a clear little trill, often repeated by both sexes, and a dry *k'wek,* apparently restricted to adult males. The trill, familiar since childhood, might guide a female ready to lay her first egg to an appropriate father for her nestlings. Moreover, I believe that the habit of courting in assemblies, which greatly facilitates the finding of males even by experienced females, should be especially helpful to virgin females. At these gatherings they would see individuals like their mothers (and themselves) seeking the courting males, and might follow their example. This appears to be an additional advantage of leks.

Lest it be thought that males who display in assemblies are invariably more ornate than the females of their species, let us glance briefly at an exception to the rule. Through a long breeding season, male Ochre-bellied Flycatchers (*Mionectes oleagineus*), stationed within hearing of each other in lower levels of tropical American rain forests, call and display in a loose courtship assembly or dispersed lek. They prepare no special courts but perch on branches of small trees. Their display consists simply of flipping their wings above their backs, one at a time. Their song is an unmusical *whip wit whip wit wit chip chip chip chip chip chip.* Like the females, they are among the plainest of birds in their unadorned greenish olive attire. They sing and display too constantly through the day to help the females build their charming, pensile, moss-covered nests with a side entrance, or to feed the young.

Like manakins, cotingas, bowerbirds, and most birds of paradise, Ochre-bellied Flycatchers are largely frugivorous. Fruits, which plants offer freely to birds who disseminate their seeds, are more readily gathered than insects, which try to hide. The ease with which fruit-eating female birds can collect enough food for themselves and their young makes them less dependent upon male help to rear their broods. Although many give their nestlings more insects than they themselves eat, by gathering fruits they can quickly satisfy their own needs, leaving ample time to search for insects and other small invertebrates for their callow dependents. Moreover, altricial birds that lack male assistance usually have very small broods, commonly only two nestlings and often, among birds of paradise and cotingas, only one, although among tropical passerines with biparental care, a clutch of two or three eggs is more usual. The frugivorous female's competence to rear her small brood without male help has emancipated the latter from parental duties and permits him to join a courtship assembly or to build bowers.

Although the males of a number of insectivorous passerines neither contract pair bonds nor help at nests, they do not join in courtship assemblies, and they rarely become as ornate as many birds that perform in leks. Among these insectivorous males that remain aloof are the Sulphur-rumped Flycatcher (*Myiobius sulphureipygius*) and related species. The sexes are alike, and as they flit through the woodland catching insects on the wing, they spread their tails and droop their wings, exposing their bright yellow rumps in a sort of perpetual display; but they lack ornamental plumes. Why the emancipated males of these and a number of other small American flycatchers have not become more ornate is puzzling. The reason may be their failure to join in courtship assemblies because their insectivory does not permit them to spend so much time in such gatherings as frugivorous birds commonly do.

Exceptional among insectivorous passerines is the rare Buff-tailed Sicklebill, an elegant bird of paradise that we earlier noticed for his inverted display that exhibits fanlike expanses of elongated feathers on the sides of his neck and breast. For more than half the year, a territorial male watched by Beehler sang and displayed high in a forest tree, with no other of his kind and sex within hearing. Apparently, the long, slender, decurved bill of this species enables the female to provision her (single?) nestling with arthropods extracted from bark and knotholes, without a mate's assistance. This manner of foraging reminds one of that of some tropical American woodcreepers, one of which, the Buff-throated Woodcreeper (*Xiphorhynchus guttatus*), has a rather similar way of courting. Taking no part in nesting, a solitary male clings high in leafy trees, persistently calling with ringing notes during the breeding season. If he has a visual display, he is so well hidden by foliage or vine tangles that I have missed it. In any case, his cryptic rufous-and-brown plumage lacks adornments. As in the above-mentioned flycatchers, freedom from parental duties has not led to ornamentation.

Frugivory favors courtship assemblies not only by easing the burden of finding enough food but also by promoting mobility. Especially in the tropics where most leks are found, insectivorous birds tend to be more territorial and stationary than frugivorous birds, which wander widely in search of trees and shrubs that ripen their fruits now here, now there, often at considerable distances. The success of a courtship assembly depends upon the number of females it can attract, and the more mobile or less confined to a small territory the latter, the greater the probability that they will visit one that is well situated. This makes participation in an assembly more profitable for frugivorous birds than it would be for insectivores. The greater the number of males who offer themselves for female approval, the greater the probability that outstanding beauty will evolve. Whether they display in compact assemblies, in exploded leks, or with greater separation, they should not be so far apart that a female's choice is narrowly limited. Males who now display far from others may owe their splendor to ancestors that in past ages were more numerous and concentrated, offering females a wider range of choice.

Fortunately, birds can become beautiful in plumage without being relieved of parental chores and forming courtship assemblies. In the following chapter, we shall examine an alternate route to avian splendor.

References

Bateson, ed. 1983; Beehler 1987a; Beehler and Foster 1988; Huxley 1938a; 1983b; LeCroy 1981; Payne 1984; Pugesek and Diem 1986; Shaw 1985; Skutch 1960; 1972; Stiles and Wolf 1979; Williams 1983.

15.
Mutual Selection in Birds

In the preceding chapters, we gave much attention to courtship assemblies and bowers. In the former, sexual selection produces its most impressive results, as in the lavish plumage of birds of paradise and the elaborate rituals of manakins. Bowers are of outstanding importance in the interpretation of avian adornments because they show conclusively that their primary function is to win females rather than to overawe rival males. Nevertheless, these modes of courtship are restricted to a small fraction of the feathered world; in only four families — hummingbirds, manakins, cotingas, and birds of paradise — are courtship assemblies frequent, while bowers are built by only a single small family. The great majority of avian species have other mating systems. Relatively few are polygynous, a male mating with two or more females, whose broods he sometimes attends, or promiscuous, the male inseminating a number of females with whom he forms no pair bond. Fewer still are polyandrous, a single female associating with several males. Approximately 90 percent of all birds breed in monogamous pairs. Many of these birds are no less beautiful than those in courtship assemblies, but their adornments tend to be more restrained, because long plumes and dangling wattles would interfere with the performance of parental services in which monogamous birds commonly participate. What factors contribute to the rich coloration of many monogamous birds?

Monogamous birds fall roughly into two great categories. In the first of these divisions, one sex, nearly always the male, chooses a territory or builds a nest before he has a mate, then tries to attract one. Such birds are usually migratory or widely wandering rather than permanently resident. I have already (Chapter 5) called attention to some of the similarities between this mating system and courtship assemblies, as well as some of the differences. Intersexual selection prevails, because a female chooses among several males in neighboring territories who try to win her as a mate. Her choice is influenced not only by the characteristics of the males but also by the qualities of their territories, whether they contain enough food, good nest sites, and the like. The male, if truly monogamous, is satisfied with a single partner. If, as frequently happens, the same male and female nest together in successive years, this is often because, after an interval of separation in the nonbreeding season, both return to the same territory where they have already successfully bred. In these birds, pairs form rather rapidly, often in a day or two, and their formation is fairly easy to watch. If the male is very brilliant, his mate tends to be plainer.

The second category of monogamous birds includes both migratory and resident species, with the latter predominating. Pairs tend to form long before the breeding season, apparently most often by mutual selection; but the process by which alliances are formed is, as a rule, much more difficult to follow than in birds of the preceding category. However, the displays that apparently help to bring the partners together are often continued after the pair has formed and may be readily observed. When resident, and sometimes even when migratory, these birds often live in pairs throughout the year; probably the same two remain together as long as both survive; but this has been proved for only a few species, and the appearance of lifelong fidelity could be maintained by frequent changes of partners who look very much alike. In birds that practice mutual selection the sexes often differ little in appearance and behavior; they wear the same colors, take equal parts in displays, and share more or less equally in attending nests and young. The lack of obvious sexual differences in birds with mutual selection frequently creates a problem for the ornithologist who tries to learn the roles of the sexes in reproduction. Even coition may be no trustworthy guide, for reversed mounting occurs. An infallible, but often time-consuming method of distinguishing the sexes without harming the birds is to see which member of a pair lays an egg.

Often birds only a few months old form pairs that will not nest for many additional months, sometimes more than a year. In Central American highlands, I found three species of wood warblers, the Slate-throated Redstart (*Myioborus miniatus*), Collared Redstart (*M. torquatus*), and Pink-headed Warbler (*Ergaticus versicolor*), whose young, hatched in April and May, had by August or September acquired adult coloration and lived in pairs that would not nest before the following March. Chestnut-capped Brush-Finches (*Atlapetes brunneinucha*) molted into adult plumage and took partners similarly early. At lower altitudes in Costa Rica, a few months after the nesting season, I rarely see a Golden-masked Tanager (*Tangara larvata*) who is not in adult attire and with a partner. These are only a few of the examples of early pair-formation that might be adduced. I have not learned how these associations arise. Among permanently resident tropical birds, who have ample time to select partners and territories, such arrangements are made much more obscurely than among migratory birds, who must make them hurriedly.

Even in the North Temperate Zone, where winters are frequently harsh, Eastern Bluebirds (*Sialia sialis*), hatched in May, sometimes have partners by September. Also in the United States, nonmigratory Wrentits (*Chamaea fasciata*) and Plain Titmice (*Parus inornatus*) pair in the fall, soon after separating from their parents. In Europe, Bearded Tits (*Panurus biarmicus*), still wearing juvenile plumage, pair when only 2.5 months old but do not breed until 9 months later. Jackdaws (*Corvus monedula*), much larger birds that develop more slowly, first choose mates at the age of about one year, twelve months before they will build nests, Royal Albatrosses (*Diomedea epomophora*) often pair as much as eighteen months before they will start to breed at the age of nine or ten years.

The interval between joining in pairs and undertaking the responsibility of parenthood is sometimes called the "engagement period," during which the partners have ample time to assure compatibility for the tasks that lie ahead. Equality of age may be important; as earlier mentioned (Chapter 14), cormorants, gulls, and other birds tend to choose partners of their own age. Precociously formed pairs may dissolve before they start to breed, perhaps in some cases because, with approaching maturity, the companions discover that both are of the same sex, which is not improbable in species without obvious sexual differences. Although this has been doubted, I have known two unmistakable domestic cockerels to keep close company. Konrad Lorenz told how two male Greylag Geese (*Anser anser*) sometimes live as a pair. In the forest of Panama, Edwin Willis knew two duos of male Bicolored Antbirds (*Gymnopithys leucaspis*) who kept company like mated pairs, often passing food back and forth. Whatever the reason, pairs of immature Eastern Bluebirds, Eurasian Blackbirds (*Turdus merula*), and Wrentits often separate before their first breeding season. Mature birds who have successfully nested together are more likely to retain the same partners.

Among the Yellow-eyed Penguins (*Megadyptes antipodes*) studied by L. E. Richdale in New Zealand, pairs were formed at any time of the year but most often in winter. Either sex might take the initiative in pair-

formation, as is to be expected when selection is mutual. The most ornate of the penguins are the six crested species of *Eudyptes*. The Rockhopper Penguin (*E. crestatus*), which breeds on sub-antarctic islands
around the globe, stands about one foot (30 centimeters) high. Both
sexes have dark gray heads, upperparts, and throats, with the remaining
ventral surface white. They have low crests of dark, upwardly curved
feathers, pale yellow eyebrows, and loose tufts of straw-colored feathers
drooping from their hindheads. Their thick bills are orange, their eyes
bright red. The mutual displays that help to keep the pair bond strong
were studied by John Warham on Macquarie Island, south of New Zealand. Mutual preening, one of the most widespread social activities of
birds of many kinds, is frequent whenever mated Rockhopper Penguins
are together and often follows their displays. The partners turn their
heads sideways and, with the tips of their bills, nibble each other's
throats and necks. When a penguin first permits an individual of the
opposite sex to join it, they exchange this courtesy. Nonbreeders who
keep company always preen their partners. Parents nibble the feathers
of their chicks, and the young birds reciprocate.

Often a penguin, bowing so deeply that its bill almost touches its feet,
utters deep, throaty, throbbing notes, its body shaking to the rhythm of
its calls. A solitary bird of either sex may perform in this manner; but
when the bowing penguin's mate is near, it usually bows, too, both bending low and calling with their heads close together. This behavior is so
contagious that a penguin can hardly resist bowing when those around
it bend low; a parent about to feed its chick may be so strongly impelled
to imitate bowing bystanders that it fails to deliver the food.

Penguins of different species often greet their mates by trumpeting.
As a Rockhopper approaches its nest to relieve its sitting partner, they
stretch their open bills toward each other and trumpet loudly. Neighbors
often join in, directing their shouts toward the newcomer. When the
latter steps into the nest, the pair assume a vertical stance, stretch up
their necks, point their widely open bills skyward, and wave their flip-
pers up and down in time with their braying. The muscles of their chests
ripple and swell as they pour forth their notes with evident gusto.

The heads of some of the grebes are elegantly adorned in the breeding
season and, correspondingly, their displays are elaborate. Alike in plumage, the sexes perform together in almost identical fashion, or else their
roles are reversed, one and then the other giving the same act. Those of
the Horned Grebe (*Podiceps auritus*), which nests in Alaska and central
and northwestern Canada, were described in detail by Robert W. Storer,
an authority on this family of diving birds. In their nuptial attire, both

Rockhopper Penguins, *Eudyptes crestatus*. Sexes alike.
Subantarctic islands.

sexes wear a small, upcurved crest at the back of the crown, tufts of buffy feathers, the "horns," on the sides of their heads, and a rounded black muffler around the top of their long necks, which below this are chestnut. On the female these adornments, which enter prominently into the grebes' displays, are only slightly smaller and paler than on the male. Some of the most striking acts follow in a fixed sequence called the "discovery ceremony," which is performed in spring when mates have been separated for a long time or by a considerable distance. This ritual is usually preceded by "advertising," when the grebe rides high in the water and stretches up its neck, with head plumes prominently spread, as though looking for its partner. Then one bird dives and the other gives the "cat display," in which the grebe, swimming high in the water, raises its wings above its back and retracts its head with spread plumes. This grebe's mate performs the "ghostly penguin" act: after the last of a series of "bouncy dives," the bird emerges with its body nearly upright but its head horizontal and turned away from its partner. As it rises higher into the air, it turns to face the partner, who also stretches upward until the two, appearing almost to stand upon the water, join in the "penguin dance." Then one or both preen the feathers of their backs and wings before they rise again in the penguin dance. The dance and preening may alternate about ten times, forming the climax of the discovery ceremony, before the participants turn formally away from each other.

Sometimes, after turning away, one Horned Grebe dives and emerges with aquatic weeds in its bill, or both partners do this. When both bring up weeds at nearly the same time, they swim rapidly side by side for from a few feet to perhaps ten yards. Sometimes they rush back and forth; at other times their courses are random. A "weed ceremony" may consist of a single rush or more than ten, usually at least four or five. After each rush the grebes separate, subside into the upright posture, swim around briefly, then rise into the penguin posture as they join for more rushing over the water. The act ends when one or both participants drop the weeds.

To describe all the ceremonies in which Horned and other formalistic grebes engage would, I fear, bore the reader, who would rather see than read about them. Nevertheless, two more are so curious that I cannot refrain from telling about them. The recently discovered Hooded Grebe (*Podiceps gallardoi*) of southern Patagonia includes in its discovery ceremony an act, called "sky jabbing," known in no other species. The two participants approach each other face to face until their protruding breasts almost touch. Exaggerating this peculiar posture, they turn their bulging breasts upward, throw back their necks to touch their backs, and point their bills skyward. In this strained attitude, they jerk their heads rapidly up and down "with a motion like that of a sewing machine upside down." This performance is accompanied by three notes repeated as a trill — a vocalization as strange as the display.

Even more extraordinary is the bumping ceremony of the White-tufted, or Rolland's, Grebe (*Podiceps rolland*), widespread in southern South America. It starts with a pair of these birds facing each other on the water, their bodies horizontal, their necks erect. Then one dives, while the other bends down its head and appears to watch it. The submerged grebe suddenly shoots up under its waiting partner with sufficient force to knock it clear of the water. Although the birds usually make contact breast to breast, the bumping bird sometimes miscalculates and hits the other with its head. The diving and bumping is usually repeated by either the same individual or its companion, occasionally six to eight or more times. Although each partner plays both roles, they do not alternate regularly. Sometimes the ceremony ends with one of the participants diving and bringing up water plants. Some birds have odd ways of demonstrating their mutual attachment.

In contrast to the generally dark plumage of grebes, many other water birds are largely white. People who enjoy color should love white, which is composed of all the colors of the spectrum, the whole range of visible light waves, reflected from the same surface and skillfully blended by our visual apparatus. One of its greatest attractions is its association with cleanness and purity, as in foods, fabrics, walls, and untrodden snow. Probably white would be more appreciated as a pleasing color if it were not so widespread. The rarity in tropical American woodlands of predominantly white birds, chiefly a few members of the cotinga family, makes seeing one of them a memorable occasion. White birds are found chiefly in open spaces — marshes, lakes, and, above all, the oceans — and most are of middle size or larger. Viewed against the verdure of marshes, the ultramarine of the sea, or the azure of the heaven, they are as lovely as white clouds in a serene sky. Their whiteness is caused by the reflection of all the Sun's visible rays from plumage devoid of pigment, which on many white birds is found chiefly toward the ends of wings and tail, in the form of black melanin, which increases the feathers' resistance to wear. Birds save materials by not needlessly coloring their plumage.

White birds include certain albatrosses, tropicbirds, boobies, pelicans, egrets, ibises, spoonbills, storks, cranes, swans, geese, gulls, and terns. All these birds form pairs, in the breeding season and often more permanently; most take rather equal shares in nest attendance and indulge in mutual displays, of which the dances of cranes are probably the most spectacular and widely known. Dale Rice and Karl Kenyon described the elaborate dance of Laysan Albatrosses (*Diomedea immutabilis*) on Midway Atoll in the Pacific Ocean. It starts when a male and female stand face to face, assume the "gawky look" posture, and gently nibble each other's bills. They bow to their partners, bobbing their foreparts down and up. They clapper their bills, opening and closing them so rapidly that the lower mandible is blurred, making a sound like the drumming of a woodpecker. With bills nearly closed, they utter a call that resembles the whinny of a horse. They turn back their heads, place the tip of the bill near the bend of an uplifted wing, and lightly clapper for a second or two. At the end of this display, each albatross stretches its head skyward and snaps its bill just once. Swinging its uplifted head from side to side with open bill, it emits a smooth, prolonged whining *wheeeeee.* Standing on tiptoe, necks stretched up in statuesque dignity with bills pointing toward the zenith and mandibles only slightly parted, it gives the sky call, a cowlike *moo* delivered only in this posture. The partners usually clapper in unison, but other components of the ceremony are not always performed simultaneously.

The Laysan Albatrosses begin these displays as soon as they arrive on their nesting ground. Breeding pairs indulge freely in them only until they start to prepare their nests; but unemployed birds continue to dance throughout the breeding season, with decreasing frequency toward its end. Laysan Albatrosses do not breed until at least seven years old, but at the age of three years they may return to the atoll, keep company in duos, dance, and sometimes build nests. Pair-formation is apparently a protracted process, but once the bond is forged, it tends to endure as long as both members survive.

More spectacular are the displays of the larger Wandering Albatross (*Diomedea exulans*). While the female stands upon the nest that she is building, her mate arrives with a billful of seaweed or some other material and, groaning, bows to her. She responds by bowing with a groan, takes the material from him, lays it on the nest, and stamps it down with her feet. Her partner sits on the ground and makes a bubbling sound in his throat. Next, throwing up his head, he brays loudly, as does his mate. They nibble the feathers of each other's heads and necks, touch bills, and rattle their mandibles. The female steps off the nest and, while the male marches solemnly around, turns to keep her face toward him. He spreads his great wings, nine or ten feet (2.7 to 3 meters) from tip to tip, lifts his head, and rattles his bill. Frequently the female does likewise. Then the two stand facing one another in a magnificent pose, while with rattling bills they repeatedly bow their heads to touch their breasts. This mutual display, often repeated early in the breeding cycle, usually culminates in coition.

Mainly white, with pale gray backs, black caps, and frequently long forked tails, terns are among the most graceful of birds. What could be more dainty and etherial, more fairylike, than the little White Tern (*Gygis alba*), with no dark feathers on its immaculate plumage except around its eyes, which balances its single white egg upon the branch of a tree, with no vestige of a supporting nest? A male tern often begins his courtship by flying rapidly upward for hundreds of feet, closely followed by a female. From the apex of this ascent the two glide earthward in a zigzag course. Later, the male flies with a fish in his bill, or the two sexes course through the air together, each carrying a fish. The partners may pass a fish back and forth between them before the female swallows it. On the ground, the male tern struts and postures before his mate at the site he has chosen for their nest. Breeding in enduring monogamous pairs, the sexes share equally all the tasks of parenthood. Although mates may separate when they migrate, they tend to reunite at the same nest site year after year.

Graceful at all seasons, pure white egrets, and herons with more varied colors, are beautiful in their nuptial attire, when long, filmy plumes flow from their heads, necks, and backs — aigrettes that caused their bearers to be slaughtered in enormous numbers to adorn women's hats. Clad in rich chestnut and dark glossy green, with a long, blue-gray crest, the Chestnut-bellied Heron (*Agamia agami*) is so elegant that my first sight of one, in a tiny Guatemalan marsh shaded by willow trees, remains vivid in memory after a lapse of over half a century. Many species of herons and egrets nest in colonies, often with ibises, anhingas, spoonbills, and other marsh birds. A male chooses a nest site, perhaps the remains of a nest from the preceding year, and by displaying and calling holds other males aloof while he tries to attract a mate. Like many another bird zealously defending a nest site or a larger territory, he resists the intrusion of any other individual, not excepting the female he needs — and this resistance is likely to be strongest in species without obvious sexual differences, as in the heron family. Females gather around him, watching his displays, hesitating to risk a jab from his sharp bill by approaching too near. They have different ways of overcoming his sharpness. A female Great Egret (*Casmerodius albus*) boldly advances, often while the male is busily arranging sticks in his nest platform, pushes beneath him with compressed plumage, retracted neck and head, and displays by snapping her bill. Despite her confident attitude, he drives her away. She returns, is again repulsed, but joins him persistently until she overcomes his resistance and wins acceptance.

The complex situation in a heronry contains elements of intrasexual selection in both sexes mingled with intersexual selection. Males compete with males for nest sites and mates. Females vie with females for nest sites or platforms. However, intersexual selection by females who choose and try to win males by overcoming their resistance, in some species forcefully, overshadows the other modes of selection. This raises a problem. Should not the females remain unadorned, as in many other birds with strong intersexual selection? Why are their colors and nuptial plumes no different from those of the males? To be sure, there is no good reason why they should not be, for in a crowded, noisy colony, cryptic coloration does not help nest attendants to remain undetected by predators, and the sexes take rather equal shares in incubating and caring for the young. A more positive factor for keeping the sexes alike appears to be the mutual displays in which they engage after a sometimes stormy pairing. Herons and egrets are ceremonious birds. Their elegant adornment in the nesting season has its counterpart in their elaborate rituals. Soon after the male has accepted a mate, the two proceed to finish the nest that he has started. As in pigeons, the female sits on the platform to receive and arrange the sticks that her partner, flying back and forth, brings and with formal gestures presents to her.

As an example of herons' displays, let us take the Tricolored, or Louisiana, Heron (*Egretta tricolor*), a slender, graceful bird handsomely arrayed in slate-blue, purple, maroon, and white, studied on the Gulf coast of Louisiana by Julian Huxley and, many years later, by James A. Rodgers, Jr. Before they start to build, the newly wedded pair enjoy a honeymoon, for long hours sitting side by side, one resting its head against its spouse's flank. From time to time, they face each other and, with loud cries, cross or intertwine their long necks, while each nibbles the aigrettes of its consort. After they begin to build, the male, returning to the nest site with a twig in his bill, stretches his head upward, extends his wings sideward, and erects his ornamental plumes. Repeatedly he lifts his bill with the twig toward the sky, then lowers it toward the nest, the while calling *culh-culh* over and over. The female on the nest acts similarly while she receives the stick from her mate. Both nod their heads again and again, call a few times more, and caress each other's bills. Then, with a vibratory movement, the female works the twig into the nest, while the male watches intently. Finally, they lay their plumes flat, and the male flies away, to bring twig after twig to his waiting partner, each time with a repetition of the greeting display.

After the herons' eggs are laid, the mates sit alternately for long intervals, while the unengaged partner forages at a distance. Returning for its spell on the eggs, the heron approaches the nest by a long glide with its neck stretched fully forward, instead of retracted in the form of an S, as in normal flight. When it reaches the nest, the two birds join in the greeting display, just as while they built. If the incubating or brooding partner delays to relinquish the nest to the newcomer, the latter continues to perform the greeting ceremony, with intense bill nibbling, until the other is persuaded to depart. Instead of flying afar, the heron just relieved of sitting finds a stick, brings it to the nest, and presents it with the greeting display to the bird now settled there. After one changeover, up to eleven sticks may be delivered in this ceremonious manner; but, with many repetitions, the displays after each presentation become more perfunctory and less intense.

Both sexes of Tricolored Herons bring sticks to the nest at the change-overs, and they may continue this activity until the growing nestlings are no longer brooded. All these sticks may not be needed to maintain the nest in good repair; but bringing them, with the accompanying mutual displays, helps to keep the pair bond firm during the weeks when the partners are separated for long hours, and may see each other seldom except when they exchange places on the nest. An incubating or brooding Tricolored Heron, returning to the nest that it has left uncovered while it preens or sunbathes nearby, often gives the greeting display in the absence of a partner. A parent arriving with a meal for hungry, rudely snapping nestlings may calm them with a greeting display. Two fledged siblings, approaching each other after an interval of separation, meet with greeting displays. These mutual demonstrations mitigate the aggressive tendencies of herons.

In American and African parrots, the sexes are alike; in some Australian parrots, they differ. Parrots tend to live in pairs throughout the year. As large flocks of Scarlet Macaws (*Ara macao*) or Red-lored Parrots (*Amazona autumnalis*) pass overhead, it is easy to see that pairs fly wing to wing, more widely separated from other pairs. Often the flock contains a few single birds, some of whom seem to try to intrude into established pairs, although they might be the pairs' grown offspring. In other, more swiftly flying flocks, such as those of Brown-hooded Parrots (*Prionopsitta haematotis*) and Sulphur-winged Parakeets (*Pyrrhura hoffmanni*), pairs are not evident. Mutual preening, duetting, feeding of the female by the male, and resting or roosting side by side or sleeping together in a cavity are widespread among parrots and help maintain the pair bond at all seasons. Mated parrots may make good use of their flair for imitating sounds by matching the voices of their partners, thereby establishing another link between the two. Male parrots often help to excavate the nest cavity in a termitary or decaying tree; they rarely incubate; but they commonly feed their incubating consort and the young, in both cases by regurgitation. Because wild parrots are usually shy and their nests, often high in dead trees of uncertain stability, are difficult or dangerous to inspect, their mating and breeding habits have been studied chiefly in aviaries, probably to a greater extent than in the case of any other family of birds.

Hilda Cínat-Tomson studied courtship and pairing of captive Budgerigars (*Melopsittacus undulatus*) by placing a female in a cage with several males and permitting her to choose among them. After experimenting with ten individuals of each sex, Cínat-Tomson concluded that the female Budgerigar consistently selected the prettiest of the males offered to her. Since the most beautiful of these little male parrots were also the most spirited, it was uncertain which of these factors more strongly influenced the females' choices. The experimenter cut from some males the black spots that ornamented their yellow collars, and later attached these same spots to other males who lacked them. When offered males with black spots removed along with males who had never had any black spots, a female was as likely to choose one as the other. But when given the choice between an originally beautiful but now shorn male and another who had been artificially embellished, the female selected the latter. Males stained to resemble females were treated as such by other Budgerigars, and vice versa. Females appeared not to distinguish between the voices of several males. Visual rather than auditory impressions played the principal role in mate selection, the females usually favoring the most handsome males.

Among passerine birds, the engaging courtship ceremony of the sociable Cedar Waxwing (*Bombycilla cedrorum*) has been described by Loren S. Putnam, Rosemary Gaymer, and others. While the birds perch in a row in a tree, one hops sideward toward another. If the waxwing approached is not receptive, perhaps because it is already mated or prefers a different partner, it gives a threat display and the active bird retreats. If the individual approached is in the proper mood, it hops a short distance away along the perch, then promptly returns to the first. The bird who starts the performance is probably a male, but this is difficult to prove, because the sexes are alike and both are songless. Often he begins by bringing a berry or other tidbit and presenting it to the female. She accepts it, hops a few inches away, then back, and returns the berry to him. He sidles away, hops back, and again passes the item to her. This alternation may be repeated again and again in a stereotyped pattern that reminds the onlooker of a spring-driven mechanical toy. A dance of a dozen hops may be interrupted long enough for the male to find another berry or insect to pass to his partner, who usually eats it as the performance ends.

Among small passerines, mutual displays tend to be less complex and spectacular than among larger birds such as grebes and penguins. In many years of studying passerine birds in tropical America, the most elaborate mutual display that I have watched was that of already-mated Black-capped Donacobiuses (*Donacobius atricapillus*), elegantly attired marsh birds widespread in South America. Formerly classified in the mockingbird family (Mimidae), they are now believed to be highly atypical wrens. While resting side by side, or clinging one above the other on an upright stalk of a tall marsh grass, the mates spread their long tails until the pattern that each presents is a wide, dark central stripe broadly bordered with white. Simultaneously, the two birds wag their fanned-out tails rhythmically from side to side through wide arcs, while with black bills widely open they voice contrasting notes. The male utters a loud, liquid, ringing *whoi-it whoi-it whoi-it . . .* , or sometimes a higher note, while his mate accompanies him with a sizzling or grating sound. One morning, in a territorial dispute, two pairs displayed to each other at short intervals. When the display was most intense, the birds' backs were humped, their tails depressed, their heads lowered, and their throats grotesquely distended, doubtless to provide resonance for the loud notes. When less highly excited, displaying donacobiuses perched more upright while they wagged their tails and called.

More often than by elaborate visual displays, passerine birds that live in pairs throughout the year keep the bond firm by their voices, frequently by duetting, which may be simultaneous or antiphonal. For birds who forage amid dense vegetation where visibility is narrowly limited, voice is indispensable for maintaining daylong contact. Wrens gleaning amid tropical foliage sing frequently and well at all seasons, often articulating their alternating phrases so skillfully that unless one stands between the singers one seems to hear the continuous song of a single bird. In a number of species, females sing in their nests, answering the song of their mates. Pair members who look alike may have quite different voices. Female Buff-rumped Warblers (*Phaeothlypis fulvicauda*) answer with low, sweet warbles the loud, jubilant, ringing crescendo of their mates. Foraging amid dense thickets, the black male Yellow-billed Cacique (*Amblycercus holosericeus*) utters a beautiful, clear, liquid whistle of two notes, which his equally black consort answers with a prolonged, rattling *churr*. This sequence of contrasting sounds may be heard throughout the year, assuring us that these birds so difficult to see remain paired at all seasons. Among nonpasserines, the loud duets of Gray-necked Wood-Rails (*Aramides cajanea*), sounding like *chirin-co-co* many times reiterated, are often heard by night as well as by day.

Just as unilateral selection, especially as it occurs in courtship assemblies, tends to make the sexes of birds very different in appearance and activities, so mutual selection tends to make them alike in both respects. Its most valuable consequences are nuptial constancy and cooperation, with a large bonus in the form of beautiful plumage and delightful songs. Darwin considered the suggestion that mutual selection may be responsible for the similar adornment of the two sexes, but he rejected this idea because male animals are so ready to mate, unselectively, with any female. In his day, it was not known that many birds select partners while they are still sexually inactive. If, as is widely conceded, males in courtship assemblies owe their beauty to the choices of females, it would seem inconsistent to deny that selection of partners by both sexes can have similar effects on both sexes. The beauty of many birds that practice mutual selection and the striking differences in the patterns and colors of related species far exceed the requirement of distinctive markings to prevent hybridization by species that occur together. Some species so similar in appearance that only experts in field identification can distinguish them consistently avoid mismating. Different vocalizations or courtship displays may help to keep these similar species apart.

It is not difficult to imagine how both sexes of monogamous birds might become equally colorful. Let us begin with a species, of which innumerable examples might be cited, of which both plainly clad parents attend the young. In due course, these young birds would select partners like their parents. In view of the widespread tendency of birds to prefer the bigger or the more colorful — the supernormal stimulus — a mutation that added a touch of color to the originally plain adult plumage, particularly to feathers erected or exposed in moments of excitement, would be attractive to both sexes of these birds. By the slow processes of mutation and selection, and the continued tendency of each generation to choose mates like their similar parents, but sometimes with a little color added, the sexes should increase equally in brilliance — unless the need for obscure secrecy prohibited brilliance in the sex, usually the female, that spends more time at the nest. This appears to be the route by which both sexes of many tropical birds, notably numerous parrots, tanagers, wood warblers, and orioles, became equally colorful.

I had not been long in tropical America before I was impressed by the fact that some families, in which the sexes of migratory species are often very different, have among their members permanently resident in the tropics many species in which the sexes are colored alike. I was at first inclined to attribute directly to the migratory habit this striking difference between migratory northern birds and permanently resident tropical birds. Now it appears that migration is responsible for this difference only indirectly, by causing a difference in the mating systems. As already noticed (Chapter 5), the mating system of many migratory birds, of which the males arrive first in spring and acquire territories to which they invite females, has much in common with courtship assemblies or leks. In both categories, unilateral sexual selection prevails, and the consequences are the same — pronounced differences in the appearance of the sexes, except when ecological factors keep both cryptically colored. In contrast to this, mutual intersexual selection, widespread among permanently resident birds, tends to make the sexes alike. There appears to be no strong selective pressure to keep incubating and brooding females duller than the males who will frequent the nest during the period of greatly increased activity when both sexes feed the nestlings.

Although the sexes of birds among which mutual selection prevails are frequently alike in appearance, they may differ greatly. This is true of the Zebra Finch (*Poephila guttata*) of Australia, a small, highly gregarious grassfinch (Estrildidae) frequently bred in aviaries and biological laboratories, where it is widely used in experiments. The male has a gray back and wings, white rump and upper tail coverts, the latter banded with black, white cheeks margined with black, chestnut ear patches, throat and chest finely barred with black and white, a broad black breast band, white abdomen, and brown sides spotted with white. His bill is bright red, his legs and feet orange. The female is largely gray, with rump, upper tail coverts, and cheek patch as in the male. Her bill is paler red than the male's; her legs and feet are orange like his. Klaus Immelmann, who studied Zebra Finches in Australia, found that they form pairs while flocking at a distance from where they will nest. Although the ornate males usually take the initiative, females occasionally do so; and, in any case, the final decision is theirs. Pairs remain together from nesting season to nesting season, apparently as long as both members live. They perch in contact, preen each other, and sleep together in a covered nest with a side entrance. Despite the prominent differences in the coloration of the sexes, Zebra Finches behave like other birds who practice mutual selection.

Nancy Burley made an elaborate experimental study of the color preferences of tractable Zebra Finches. In each trial, a single individual was permitted to perch beside any one of four others, one of whom remained unbanded, while the others wore leg bands of different colors, each with the same color on both legs. When females were permitted to choose between four males, three of whom were banded with, respectively, red, orange, and green, while the fourth was bandless, they favored red more than orange, orange and unbanded about the same, and green least of all. In the reverse experiment, males, given the choice between black, orange, unbanded, and light blue females, spent most time resting beside black females, least with blue, and did not discriminate between orange and unbanded. Males offered the choice between other males that were, respectively, unbanded and banded with green, orange, and red, were attracted most strongly to green-banded birds, least to those with red bands. Females tested with other females preferred blue to orange, orange to unbanded, and black least of all. Other experiments showed that yellow-banded males were highly attractive to females; but males with yellow rings on legs painted dark orange-red were favored above those with yellow bands on unpainted legs.

Burley's experiments teach us much. In the first place, they reveal how quickly some birds recognize, and respond to, relatively slight changes in the appearance of their companions. They demonstrate the value, in attracting partners, of the colorful legs of many birds, and of bills that in adults may be permanently red, orange, or yellow, or may acquire one of these colors as the breeding season approaches. They show that female Zebra Finches, whose males are more highly colored than themselves, prefer males with added color, whereas males prefer females without bright colors. Moreover, the tests reveal that the colors which attract a bird of either sex to another of the same sex are different from those which attract birds to individuals of the opposite sex. The different color preferences of the male and female Zebra Finches in these experiments accord with the different coloration of the sexes. I surmise that similar tests made with a species whose sexes are alike would reveal that males and females have the same preferences. Incidentally, the experiments make it appear probable that the increasing number of people who band free birds, especially with colored rings, alter their chances of winning mates.

We sometimes assume that birds are genetically predisposed to prefer mates with the normal coloration of their species or race, and that mutations that alter the appearance of one sex would be selected against unless supported by appropriate mutations in the preference of the other sex. Certain observations, such as those by Cherry Kearton on the persecution by flock mates of individual Jackass Penguins (*Spheniscus demersus*) that differed conspicuously from the norm, lend support to this assumption. However, it appears from these experiments that, if not too great, alterations may be not only accepted but preferred. Burley's female Zebra Finches were predisposed to prefer a slight intensification of the males' colors, which might be regarded as a positive response to a supernormal stimulus, comparable to that which prompts certain birds to incubate eggs that are larger, or more heavily spotted, than their own. When birds are preadapted to prefer enhanced coloration or ornamentation, the evolution of such embellishments should proceed much more

Hooded Warbler, *Wilsonia citrina,* male. Eastern United States and southeastern Canada; in winter, from Mexico to Panama.

wide, but because 21 percent of them sing exceptionally well. Wrens do even better, with 24 percent of 59 species in the list. However, the family with the highest proportion of superb singers is the thrashers and mockingbirds; 35 percent of its 31 species are represented. In most families, except some very small ones, less than 15 percent of the species are judged to be superior; and many others have not a single species deemed worthy of inclusion. Nevertheless, I could name many birds whose songs, perhaps with no high degree of technical excellence, I love to hear. To me, the great charm of many bird songs is not their complexity so much as their sweet simplicity. Like the timid voice of a little child, they may stir us more deeply than a brilliant performance can do.

Many songbirds have large repertoires. Individual Marsh Wrens (*Cistothorus palustris*) in the state of Washington sang more than 100 different songs, most of which were shared by all six of the males in a locality, who collectively had 127 different songs, as recorded by Jared Verner. Songs of the Sedge Warbler (*Acrocephalus schoenobaenus*), often up to a minute long, are composed from a repertoire of 50 syllable types, so arranged that no two songs are alike. Clive K. Catchpole, who studied these songs, regarded them as "the acoustic equivalent of the Peacock's train, an extravagance whose only possible function could be to influence female choice." Related species have even larger repertoires: Marsh Warblers (*A. palustris*) sing from 80 to 100 different songs; Reed Warblers (*A. scirpaceus*), 70 to 90; Moustached Warblers (*Lusciniola melanopogon*), 60 to 80. These three species of Old World warblers sing continuously for long periods.

The smaller division of the great Passeriform order, the Suboscines, is behaviorally quite similar to the songbirds. Members of this mainly Neotropical assemblage show the same range of habitats and foraging habits as do the songbirds, their nests are equally diverse, and they take as good care of their young, but their songs are simpler, produced by less complex vocal organs. With the notable exception of the lyrebirds, their songs tend to be repetitions of similar notes, or at best a few different notes, forming a trill when the delivery is rapid and the notes soft and clear, a rattle when they are dull, wooden, or harsh. Or else the song consists of a series of more or less separated, similar whistles. Variety is achieved by changes in volume and tempo. Despite their limitations, these utterances often have a ringing, cheerful quality that makes them most pleasing. The song of the Scaly-throated Leaftosser (*Sclerurus guatemalensis*), an ovenbird, is a rapid flow of pure, clear notes, delightful to hear as it issues from the dark undergrowth of tropical rain forest.

Birds poorly endowed with song often perform chiefly at daybreak, as is especially true of the great family of American, or tyrant, flycatchers. Their dawn songs, rapidly repeated for a long interval between the day's first dim light and sunrise, are amazingly diverse: loud and harsh, clear and ringing, low and shrinking, quaint, rarely melodious. The most beautiful that I have heard are the crepuscular songs of the Streaked Flycatcher (*Myiodynastes maculatus*) and related species. A few flycatchers, such as the Streaked and the Eastern Wood-Pewee (*Contopus virens*), sing again in the evening twilight; but dawn songs are rarely heard in full daylight except when the birds are highly excited. The function of these special songs is not clear. The fact that a dispute over territory at almost any hour of the day incites these birds to sing suggests that dawn songs are proclamations of territory. But why do not territorial flycatchers, like territorial songbirds, continue to assert their possession while the Sun is high? Since dawn songs are commonly delivered by birds already paired, they appear not to be used to attract mates. Could it be that they are simply outpourings of exuberant energy after a restful night, before the day has become bright enough, and insects active enough, for efficient foraging? (In *A Bird Watcher's Adventures in Tropical America,* I devote a whole chapter to dawn songs.)

Few nonpasserine birds are tuneful singers. Scarcely any birds use their voices more persistently than hummingbirds in their singing assemblies. Often their slight songs are monotonously squeaky; sometimes they have an engaging rhythm; and occasionally their performances deserve to be called songs in more than a technical sense. In Chapter 7 I told of the Wine-throated Hummingbird's song, which if delivered in a stronger voice might win him fame. In another nonpasserine family, the song of the Rufous-tailed Jacamar (*Galbula ruficauda*), composed of little squeals, rapid trills, and high-pitched whistles, attains a complexity worthy of a songbird. Deep, soft, full, and varied, the notes of the Resplendent Quetzal (*Pharomachrus mocinno*) are a fit counterpart of his glittering plumage. After emerging at daybreak from the burrow in which they sleep together throughout the year, a pair of Blue-throated Green Motmots (*Aspatha gularis*) join their voices in a duet of deliciously mellow notes, full, round, and clear. Few avian utterances move the listener so profoundly as the far-carrying organ notes which, as day ends, Great Tinamous (*Tinamus major*) send through the darkening rain forest. If the true function of art is to stir emotion, and the highest art is that which arouses the strongest, purest feeling by the simplest means, then these terrestrial birds classified as primitive are supreme artists. These few examples suffice to show that, scattered through the nonpasserine families, are a number of species that contribute substantially to nature's beautiful sounds, by the purity and richness of their tones more often than the complexity of their songs.

Another difference between songbirds and others is that, while still in the nest, young of the latter often repeat, in weaker voices, the songs of their parents, as I have never heard nestling songbirds do. While waiting for their parents to bring food, precocious Rufous-tailed Jacamars and Streaked-headed Woodcreepers (*Lepidocolaptes souleyetii*) practice the trills of adults. Before they fly from their hanging nests, Common Tody-Flycatchers (*Todirostrum cinereum*) repeat little trilling chirps, much like the trills of their parents, but weaker. Violaceous Trogon (*Trogon violaceus*) nestlings, looking through the doorway in the side of the wasps' nest in which they are raised, repeat a low *cow cow cow cow* similar to the adults' song, but softer and more subdued. These are only a few of the non-Oscine nestlings that I have heard practicing the songs of their parents. Apparently, the songs of many non-Oscine birds are innate but may have a learned component, as do the songs of many songbirds and hummingbirds.

It is widely recognized that male birds sing mainly to proclaim possession of territory and their determination to defend it, and to attract females. However, birds of both sexes sing in many other contexts, and no interpretation of bird song that neglected them could be complete. Although, as is prudent where predators abound, most birds incubate and brood nestlings in silence, not a few adults sing in the nest. In the mountains of Costa Rica, a male Rufous-browed Peppershrike (*Cyclarhis gujanensis*) sang while sitting, on one occasion repeating his loud, clear verses at intervals of a few seconds for nearly two hours. In a related family, the vireos, males of the Red-eyed (*Vireo olivaceus*), Warbling (*V. gilvus*), Hutton's (*V. huttoni*), Bell's (*V. bellii*), and Philadelphia (*V. philadelphicus*) sing while incubating or brooding; and in the last mentioned species several observers have also heard females sing in their nests. Likewise, both sexes of the Rose-breasted (*Pheucticus ludovicianus*) and Black-headed (*P. melanocephalus*) grosbeaks sing in their nests, so loudly that they help naturalists to find them. Among birds of which only the females incubate and brood, I have heard Gray-breasted Wood-Wrens (*Henicorhina leucophrys*), Yellow-tailed Orioles, Melodious Blackbirds, Scarlet-rumped Caciques (*Cacicus uropygialis*), and Blue-black Grosbeaks (*Cyanocompsa cyanoides*) sing while sitting, usually in response to their songful mates. These females appeared to be duetting, one of the activities that help to keep pair bonds firm, as mentioned in Chapter 15.

Cedar Waxwings (*Bombycilla cedrorum*) warble while they incubate. Gray-cheeked Thrushes (*Catharus minimus*) and Ruddy-capped Nightingale-Thrushes (*C. frantzii*) sing in an undertone while at their nests. Females of several species of American flycatchers, including the Lesser Elaenia (*Elaenia chiriquensis*), Common Tody-Flycatcher, Piratic Flycatcher (*Legatus leucophaius*), Gray-capped Flycatcher (*Myiozetetes granadensis*), and Vermilion-crowned Flycatcher (*M. similis*), repeat while incubating quaint little songs, heard only in association with the nest, which seem expressions of quiet contentment. Sometimes, too, they call out more loudly, answering mates, who never incubate but will later help to feed the nestlings. Loud singing while incubating appears to be an imprudent manifestation of irrepressible songfulness; subdued singing, an expression of parental emotion.

Most singing birds prefer to perform alone, each upon his chosen stage, where he is the center of attraction and intolerant of competition, especially by others of his own kind. Indeed, this is the central thesis of the theory of territorialism. Although most birds are soloists, many are choristers; and some are both by turns. Perhaps the most generally familiar of the social singers is the European Starling (*Sturnus vulgaris*), so widely distributed over Earth by humans. In North America's bleak midwinter, when trees are bare and few birds sing, it is pleasant to hear a cheerful party of these speckled birds, gathered in the crown of a leafless tree in the brief midday sunshine, entertaining themselves with a quiet, conversational medley of chucklings, squeaks, and varied whistles, perhaps more chatter than song. Rarely do their versatile voices emit a felicitous note in this inclement season, yet they so enliven the sere winter landscape that one is likely to forget that they are undesirable aliens.

At a later date, when the first suggestion of tender green tinctures long-naked trees, American Goldfinches (*Carduelis tristis*) gather in joyous flocks amid the burgeoning boughs, to join their small voices in a chorus of low, varied whistles and sprightly twitters. The habit of singing in flocks is widespread among goldfinches and siskins, even in the breeding season. T. A. Coward saw and heard a little party of European Goldfinches (*C. carduelis*) "singing delightfully whilst young, hard by, were still in the nest." In the highlands of Guatemala, early in the year when peach trees blossom and burgeoning oaks and alders are tasseled with long, pollen-shedding catkins, Black-capped Siskins (*C. atriceps*) gather among the nearly naked boughs of some budding tree and sing in concert, in the manner of their relatives of more northern lands. Their medley contains varied warbles and buzzy, insectlike notes, punctuated by questioning notes much like the *be bee* of the American Goldfinch — not an inspired performance, yet appreciated for its gladsome warmth. In the same region, later in the season, when they have apparently finished nesting, gatherings of clearer yellow Black-headed Siskins (*C. notata*) sing animated verses so rapidly that they seem to be in a hurry to finish singing and move onward. In the mountains of Costa Rica in January, a large flock of Yellow-bellied Siskins (*C. xanthogastra*) rested in the crowns of tall trees, singing and chattering through much of the bright day in a pleasant but far from brilliant chorus.

Orioles and related birds sing together at their roosts. Many Melodious Blackbirds slept in a thicket of young wild canes beside the Río Morjá in Guatemala. Arriving in late afternoon, they settled amid the giant grasses and repeated clear, melodious whistles until darkness fell. As the birds grew drowsy, their notes became lower and less frequent, until in the gathering dusk they fell asleep, as though lulled by their own soothing voices, and were heard no more until dawn. Early in the year, many wintering Orchard Orioles (*Icterus spurius*) roosted in a neighboring part of the cane brake. When they awoke at dawn, the males joined their voices in a chorus of low, rapidly warbled whistles delightful to hear. No bird that I know is more songful in its winter home.

Very different from any of the foregoing choruses is that of White-fronted Nunbirds (*Monasa morphoeus*), who inhabit lowland rain forests from Honduras to Peru, Bolivia, and Brazil. Their bright orange bills contrast rather incongruously with their somber, deep gray plumage, as though a black-clad nun colored her lips vermilion. From three to ten of these eleven-inch (23-centimeter) puffbirds line up, at intervals of a few inches, on a slender horizontal branch, or a vine stretched horizontally between two trees, at mid-height of the forest. Tilting their heads upward, the dusky choristers shout all together in loud, ringing, almost soprano voices, so vehemently that their bodies shake. For fifteen or twenty minutes, with only the briefest intervals of silence, the chorus swells louder or wanes, as more or fewer birds join in. Nunbirds breed cooperatively, up to six individuals attending a burrow in the forest floor; the largest singing assemblies are evidently formed by the temporary union of at least two breeding pairs with their helpers. Like the duets of many constantly mated birds, the nunbirds' concerts probably help to tighten social bonds. Although hardly musical, like the surging choruses of chachalacas (*Ortalis* spp.), they are such spirited performances and the participants seem so greatly to enjoy them that to one attuned to nature's sounds they are as animating as some grand concert, and in this sense they are beautiful.

To sing, birds often mount to a level higher than those at which they forage or nest, thereby increasing the range of their voices. If no tree is available, they may sing while they fly, often rising higher than the tallest tree. Flight songs are most common among birds of grasslands, marshes, moors, and other places where trees are rare or absent. Singing in flight adds the attraction of rapid motion to that of sound, making the singer highly conspicuous. Song that floats down from on high gains an ethereal quality that stirs human emotions and inspires poets. Few birds have been so celebrated in verse as the Eurasian Skylark (*Alauda arvensis*). Breaking into song when only a few feet above the ground, the plainly attired bird rises up and up until only a speck in the blue, circles around pouring his animated music Earthward, and continues while he plummets down, "true to the kindred points of heaven and home."

In grasslands in the middle of the North American continent, the black, white-winged Lark Bunting (*Calamospiza melanocorys*) sings in flight as well as while perched. In the same region, McCown's Longspur (*Calcarius mccownii*) rises slantingly upward with rapidly beating wings to a height of about fifteen to thirty feet (4.5 to 9 meters), then stretches his white-lined wings outward and upward "to float slowly down to Earth like a parachute made buoyant with music." His clear, sweet tones have won the highest praise. A neighbor of McCown's Longspur, the

Chestnut-collared Longspur (*C. ornatus*) likewise sings in flight, as does the Lapland Longspur (*C. lapponicus*) on the treeless Arctic tundra.

On the grasslands of the mid-continent, Sprague's Pipit (*Anthus spragueii*) and the much more widespread Horned Lark (*Eremophila alpestris*) fly much higher than the longspurs, the lark sometimes up to about eight hundred feet (244 meters). He climbs silently, flies in wide circles while he sings, then tightly folds his wings to drop abruptly to the ground. The pipit likewise ascends until he becomes a hardly visible speck but remains aloft much longer, sometimes for fifteen to twenty-five minutes, rising and falling as he circles widely. Silent while he falls, he repeats wonderfully beautiful strains while he rises. Finally, he plummets downward like the lark, spreading his wings just in time to avoid dashing against the ground.

Many nonpasserine birds rise into the air to deliver songs that are sometimes stirringly beautiful, although rarely as complex as those of the best passerine singers. Shorebirds that live in treeless places often sing enchantingly on the wing. Viscount Grey of Fallodon wrote that "of all bird songs or sounds known to me there is none that I would prefer to the spring notes of the Curlew [*Numenius arquata*] . . . As a rule the wonderful notes are uttered on the wing and are the accompaniment of a graceful flight that has motions of evident pleasure." He also wrote feelingly of the spring flight and note of the peewit or Northern Lapwing (*Vanellus vanellus*) "accompanied by cries of joy." The flight notes of the Common Redshank (*Tringa totanus*) also received his encomiums. Richard Vaughan found the loud, sweet, whistling flight song of the Greater Golden-Plover (*Pluvialis apricaria*) — *ter wee-er, ter wee-er* — "haunting, almost melancholic, nostalgic." When two or three of these plovers fly and sing together, "the moors resound to the most beautiful chorus imaginable of soft, melodious whistles . . . It is one of the wildest, loveliest, loneliest of all bird songs."

The Black-tailed Godwit (*Limosa limosa*) is a Eurasian bird that wanders to North America. The male's song-flight was described by Julian S. Huxley and F. A. Montague. With rapidly beating wings, the bird rises at a steep angle, repeating a loud trisyllabic call, *tur-ee-tur*. At a height of about 150 to 200 feet (46 to 61 meters), the character of his flight changes, and a disyllable replaces the trisyllable as he flies horizontally in wide circles with clipping wing-strokes. His tail, fully spread, is tilted now to one side, now to the other, causing his body to roll from side to side. The most spectacular part of the ceremony is the godwit's descent. Rolling and calls cease simultaneously as he nose-dives to the ground with wings and tail almost closed. Or, with wings about two-thirds open, he may swoop downward with the air roaring through his feathers, the sound audible at a great distance.

Unlike owls, nightjars, potoos, and a few other nonpasserines that are active only or chiefly by night, and others that are active at almost any hour of the twenty-four, passerines confine their activities to the daytime, the chief exceptions being the nocturnal migration of some of them, and the nocturnal singing of a few. Most renowned of the nocturnal singers is the Nightingale (*Luscinia megarhynchos*), thanks to its wide distribution in the Old World and the ability of its wonderfully varied song to inspire the verses of poets, who have sometimes misinterpreted it absurdly. Influenced by the old Greek legend of Progne and her sister Philomela, who after tragic experiences were transformed, respectively, into a swallow and a nightingale, some poets have fancied the Nightingale's song to be a melodious dirge. The nighttime singing of the Northern Mockingbird (*Mimus polyglottos*) has also inspired poetry, notably Walt Whitman's "Out of the Cradle Endlessly Rocking." The Marsh Wren and the Sedge Wren (*Cistothorus platensis*) frequently sing at night, as do the Reed Warbler, Sedge Warbler, Swamp Sparrow (*Melospiza georgiana*), Henslow's Sparrow (*Ammodramus henslowii*), and the Lark Sparrow (*Chondestes grammacus*). Apparently, residence amid reeds, sedges, cattails, and other low, dense vegetation of marshy places favors nocturnal singing, probably because in these open spaces nights are less dark than amid forests and thickets.

When nuptial zeal is strongest, birds who are not nocturnal singers often awake to start singing at the first suggestion of dawn, or even earlier by the light of the moon, as I have heard Rufous-collared Thrushes (*Turdus rufitorques*) do among pine and cypress trees on high Guatemalan mountains. A song ringing out in the comparative silence of night is more effective than the same song would be by day when, in the nesting season, many voices fill the air. The Nightingale would not be half so famous if he did not sing at night as well as by day. One wonders whether, when birds sing at nighttime, their mates awake to listen to, and perhaps be reassured by, them.

The repertoire of birds is usually limited. Although careful analysis, as with sonograms, may disclose many variations in tempo, tone, sequence, and intensity of notes, the songs of an individual or a population usually conform to a recognizable type, as characteristic of the species as its call notes, plumage, or the structure of its nest. A minority of birds vocalize without fixed patterns, with recitals endlessly varied. They may appear to improvise, combining unexpected notes in fresh sequences, or

Northern Mockingbird, *Mimus polyglottus*. Sexes alike.
Southernmost Canada, United States except northwest,
Mexico, Bahamas, Greater Antilles.

they may unmistakably reproduce sounds that they hear. In the former
case, we say that they sing medleys; in the latter, we call them mimics
or mockingbirds. The species that one person praises as an excellent
mimic may by another be classed as a medley-singer. The different ap-
praisals may be due to the different virtuosity of the individual birds that
each person happens to hear, or to the hearer's estimate of the birds'
apparent imitations. Species that improvise or sing medleys are often
closely related to mimics.

Some songbirds, of which the Chaffinches (*Fringilla coelebs*) studied by William Thorpe are a good example, have what we might call an innate skeleton of their species' song, which is covered with flesh and perfected by imitating others of their kind. Lacking instructors of their own species, they may fill out their songs with elements taken from other species that they happen to hear. Their repertoires are formed during a sensitive period in early life, often no longer than their first year, after which all their songs will conform to patterns that they have learned in youth.

Possibly no other birds collect such a large repertoire of imitations as do the Marsh Warblers studied by Françoise Dowsett-Lemaire. Starting during their first summer in their natal Europe, the young warblers continue to enrich their collection of borrowed songs after arrival in their winter home in Africa south of the Sahara. When they return in the northern spring for their first breeding season, individual yearlings can repeat from 63 to 84 imitated verses. An analysis of 30 tape-recorded sequences by different individuals revealed a total of 212 imitations, including those of 113 African species and 99 European species. Of the African birds, 33 were nonpasserines and 80 passerines. The male Marsh Warbler appears to be unique in forming his repertoire during sojourns on two continents. The warbler's sensitive period ends before his first return to Europe at the age of ten or eleven months; apparently, he picks up no more alien songs; but with his rich repertoire of borrowed phrases, and none that appears to be peculiarly his own, he can sing for an hour or more without a pause, altering the order of his notes to yield an unlimited variety of songs. In contrast to these warblers, other mimics, of which parrots and Northern Mockingbirds are good examples, continue indefinitely to imitate sounds that they hear. Vocally, these birds never mature, but retain to an advanced age the juvenile bird's capacity to augment its vocabulary.

In earlier chapters, we gave attention to the vocal pyrotechnics of bowerbirds and lyrebirds, both of the Australasian region. Now let us glance at some New World birds with versatile voices. A familiar example of temperate North America is the Gray Catbird (*Dumetella carolinensis*), who has, in small measure, the mimetic ability of his relatives, the mockingbirds. He has been heard to imitate not only the calls and parts of songs of such of his bird neighbors as the Northern House-Wren, Wood Thrush (*Hylocichla mustelina*), American Robin (*Turdus migratorius*), and Whip-poor-will (*Caprimulgus vociferus*), but even the yowling of a cat and the croaking of a frog. Catbirds to whom I have listened were more remarkable for the seemingly unlimited variety of the music — and much that was not music — that they poured forth in long-continued, scarcely interrupted streams than for their mimicry. As in many birds lacking definite song-patterns, the catbirds' performances differ amazingly in quality. Some individuals introduce so many harsh, grating, discordant notes into their medleys that, although amusing, they hardly soothe the ear; others sing charmingly, rarely interpolating jarring sounds into a continuous sequence of low, sweet notes. Among human musicians, we would attribute such variations to the good, mediocre, or bad taste of the composer. Might it not be the same with birds?

The Brown Thrasher (*Toxostoma rufum*), a more spirited and usually abler musician than the Gray Catbird, does not so often mar his songs with harsh or rasping sounds. Whereas the gray minstrel often sings in a low voice amid a thicket, his brown cousin rises to the treetops and sends forth his ringing notes for all the world to hear. He has the same great variety of verses, each of which he usually sings in pairs. But, to my mind, both of these northern birds are excelled by a singer who has received far less publicity, the Blue-and-White Mockingbird (*Melanotis hypoleucus*) of northern Central America — at least when he performs in his best style, for, like many a versatile artist, he is capricious and does not consistently maintain the same high standard. He is as unpredictable in his choice of a singing perch as in his choice of a theme; now he recites in an impenetrable thicket, now at the top of a lofty pine. Let us listen to him while he sings on this high station, early on a morning at the beginning of the rainy season in May, while his mate warms her two blue eggs in a blackberry tangle far below him. Now, in his glad auroral mood, he is at his best, not as an imitator but as an originator; with such bright verses of his own, why should he borrow from others? Like the Yellow-tailed Oriole, he has an abundance of short musical phrases, each of which he repeats over and over until he tires of it, then turns to another. He demonstrates the great range of his voice by alternating full, mellow whistles with light, tinkling trills. Rarely, in the freshness of morning, will he condescend to utter a borrowed phrase, such as the Whip-poor-will's call.

But who can maintain indefinitely the highest level of excellence? The effort is exhausting; genius needs recreation. When, after an hour or two, our blue-and-white musician tires of the classic style, he turns to nonsense songs for amusement. Now he interjects harsh and churring notes into his medley, and through the rest of the day will mix much chaff and patter with his golden grains of sound. After the inspired frenzy of the nesting season has exhausted itself, the blue mockingbird diverts himself with a bewildering potpourri of whistles, shrill squeals, guttural croaks, peeps, picarian churrs, screeches, and warbles, but he rarely sings in his best manner.

Those superlative musicians, the thrushes, appear rarely to copy other birds. In the British Isles, the Song Thrush (*Turdus ericetorum*) is known to imitate short phrases of a number of other species, including certain notes of the Nightingale. In the Americas, I have listened to thrushes, solitaires, and nightingale-thrushes without detecting a trace of mimicry, except in the Black Thrush (*Turdus infuscatus*) of the mountains of Mexico and northern Central America. In appearance he is almost the exact counterpart of the Eurasian Blackbird (*T. merula*): the same dusky plumage, the same bright yellow bill; but, unlike those of the Blackbird, his legs are also yellow. In voice, the Black Thrush is a mockingbird rather than a thrush — probably the best mimic in Central America. His proper song contains notes so smooth and mellow that I never tired of hearing them; but among his own incomparable verses this erratic genius intercalates many borrowed from neighbors less vocally gifted, producing a bewildering hodgepodge of sounds. The harsh, mewing notes of the blue-crested Steller's Jay (*Cyanocitta stelleri*), the call of the Whip-poor-will, the warbling of the Eastern Bluebird, the spirited *wake-up* and peculiar rattling flight call of the Gray Silky-flycatcher (*Ptilogonys cinereus*) are rendered with flawless fidelity; but the inimitable wild piping of the Brown-backed Solitaire (*Myadestes obscurus*) seems to baffle even his rare mimetic skill. All these various phrases, original and plagiarized, are liberally punctuated by a bizarre assortment of peeps, chucks, and whistles, many of which are apparently calls and flight notes of his neighbors. Some Black Thrushes have a vocal exercise that consists of running rapidly up the musical scale in a series of loud whistles, a feat that exhibits the great range of their voices, yet after all is but a copybook exercise, without a trace of the deep and tender feeling that the songs of thrushes are capable of expressing.

The foregoing birds are a small sample of the world's avian mimics, but sufficient for our purposes. Their melanges are rarely as beautiful as the pure strains of many birds who sing more coherent songs proper to themselves, nor do they stir the human listener's spirit so profoundly. We admire the range and fluency of the mimic's voice more than his taste; he entertains, and challenges us to test our own bird lore by identifying the originals of his copies. However, the real importance of vocal mimicry lies in another direction. It is evidence that birds take an alert interest in the sounds they hear, including many that appear to be wholly unrelated to their basic vital needs. Moreover, it shows that their behavior is not always strictly controlled by their genes. By choosing to copy this sound or that, by singing notes, borrowed or their own, in varying sequences, they demonstrate that they enjoy a measure of freedom. They are not mechanisms strictly governed by their heredity; their behavior is influenced by individual preferences. I have given so much attention to mimicry not only because of its relevance to the aesthetic sensibility of birds but also because the capacity for choice that it reveals appears again in their selection of sexual partners.

Birds appear to enjoy singing, which they do not only to announce their possession of territory and to attract mates but at times when it serves neither of these ends, as when they sing all together in a flock, or in a communal roost. They seem to delight in hearing themselves, and probably also their neighbors. When a versatile singer invents a new tune, he repeats it over and over; his neighbors may copy it. Jays who lack loud songs sometimes rest in solitude and continue for minutes on end to sing pleasantly in an undertone. Such *sotto voce* medleys appear to lack social or biologic significance; the jay sings for his own comfort or enjoyment, as a human hums a tune when alone, and ceases if another approaches. I would go so far as to assert that, if birds take no pleasure in singing, they are incapable of enjoyment, and that, if they find no joys or satisfactions in their lives, all their efforts to survive and reproduce are barren. They might as well be dead.

However much birds might enjoy their singing, this alone cannot account for the evolution of song. They might transmit to their progeny, from generation to generation, whatever sounds they could produce, but with little improvement of the vocal organs that determine the quality of their voices. For the evolution of the complex syrinx of passerines, some mode of selection was indispensable. Unless an improved vocal apparatus and the more pleasing notes that it produces confer an advantage in reproduction, voice could hardly evolve. Many female birds sing almost as well as their mates. Those who do not may be no less attentive to birds' notes than males are, just as people who can neither sing nor play a musical instrument may nevertheless enjoy music. Females' preference for the best singers could contribute powerfully to the evolution of bird song. Evidence in support of this theoretical conclusion has recently begun to accumulate.

It was mentioned in Chapter 11 that female Satin Bowerbirds choose males who mimic most competently. Clive Catchpole learned that male Sedge Warblers with the most complex songs win females earlier than their rivals with smaller repertoires. After acquiring a partner, male Sedge Warblers cease to sing, and they remain silent unless they lose their mates. They defend their territories by visual threat displays and

active aggression. Moreover, they find much of the food for themselves
and their families beyond their territories. These warblers sing to attract
mates rather than to proclaim possession of territories; and the choices
of females are influenced less by the quality of the males' domains than
by the quality of the birds themselves, as revealed by their songs. All the
evidence points to the conclusion that the unusually variable and complex songs of Sedge Warblers have been promoted by intersexual
selection.

In contrast to the large repertoires of the three monogamous European species of *Acrocephalus* mentioned earlier in this chapter, two
partly polygynous species, the Great Reed Warbler (*A. arundinaceus*) and
the Aquatic Warbler (*A. paludicola*), have much poorer repertoires, of
only ten to twenty songs. This appears to be because the females, uncertain of male assistance in rearing their broods, pay more attention to the
productivity of the territory that a male offers than to his personal quality and his songs. In contrast to this, polygynous Marsh Wrens have very
large repertoires. The two factors that appear most strongly to influence
a female's choice, quality of the male and adequacy of his territory, have
different effects in different species. In Northern Mockingbirds, R. D.
Howard learned that both factors have weight: females prefer males with
large repertoires, but the size of their territories more strongly influences
their choice.

The earliest sounds of birds, as of their reptilian ancestors, were probably often harsh or hissing notes, used to threaten a rival, attract a sexual
partner, or in similar social contexts. By selecting, generation after generation, males whose voices were a little more attractive, female birds
have fostered the evolution of superior vocal organs and the sounds they
make. Melodious song, like beautiful plumage, appears to be a product
of intersexual selection, often supplemented by the male birds' efforts to
improve their repertoires. Female birds, often so quiet and self-effacing,
have powerfully influenced the course of avian evolution and contributed vastly to the beauty of birds, making them attractive not only to
other individuals of their species but to ourselves, who appeared on
Earth long ages after birds arose.

In addition to the high aesthetic value of bird song, it might be said to
have moral value. Birds often settle their disputes by their voices instead
of by fighting. One method is countersinging. Hearing the song of a male
in an adjoining territory, the resident male copies it closely; the two sing
alternately back and forth with the same verses, and, probably recognizing that they are evenly matched, refrain from attacking one another.
The rarity of vicious fighting among songbirds, especially those con-

stantly resident in mild climates, no less than their beauty and melody, makes them supremely attractive to a thoughtful watcher.

References

Armstrong 1963; Bent 1942; 1950; 1953; Bent et al. 1968; Catchpole 1980; 1985; Coward 1928; Dowsett-Lemaire 1979; Grey 1927; Hartshorne 1973; Howard 1974; Hudson 1920; Huxley and Montague 1926; Skutch 1954; 1960; 1967; 1972; 1977; 1983a; 1987b; Thorpe 1958; Vaughan 1980; Verner 1976. Quote on p. 238 from William Wordsworth, "To a Skylark."

17.
Butterflies

No other division of the animal kingdom contributes so much to nature's beauty as birds and butterflies. Which contributes more is a debatable question. The advocate for butterflies might remind us that butterflies have more species than birds — ten to fifteen thousand butterflies as against about nine thousand birds. On the basic ecological principle that small creatures tend to be more abundant than larger creatures, we may confidently assert that butterflies far exceed birds in the number of individuals. Although lacking on the oceans, except as stragglers or migrants blown from their courses, on land they are found nearly everywhere that birds occur, from tropical forests to boreal tundras and as high on mountains as plants flower.

And what resplendent creatures many butterflies are! The largest birdwings (genus *Ornithoptera*) of southeastern Asia and islands of the southwestern Pacific and the morphos of tropical America have wingspans comparable to those of small birds, largely shimmering green in some of the former, the most intense azure on males of the latter. From these giants of the lepidopteran world butterflies range downward in size to miniature gems hardly an inch across. Butterflies display every bright color from red to violet, in large expanses or in the most intricate and charming patterns. Moreover, although some butterflies are wary and elusive, many are easy to approach, so that one can enjoy their loveliness at arm's length as they sip nectar from bright flowers, which is a great advantage. They do not hide their beauty but spread it to the sunshine. You do not need a binocular to see butterflies well, as you do with most birds. Not only adult butterflies are beautiful; many caterpillars are attractively colored, and even the pupal cases of some, such as the Monarch's green chrysalis, spotted with gold.

Probably the defender of birds' primacy would be forced to concede that butterflies offer as much to the eye as birds do, but one might point out that with their songs birds contribute another kind of beauty of which butterflies are devoid; a few of the latter produce a rattle; most are silent. In winter, when butterflies are dead or dormant, birds add touches of color and animation to drear landscapes just at the time when it is most needed to cheer us. Moreover, birds build charming nests and lay eggs that are often beautiful. All told, birds add more to Earth's beauty than any other animals; butterflies come second.

Birds bear their bright colors chiefly on their bodies; butterflies, almost exclusively on their wings. When we recall that other insects — bumblebees, dragonflies, burly beetles — fly very well with wings much smaller in relation to their bodies, we ask why butterflies' wings are so disproportionately expanded, so brightly colored. Many butterflies seem to invite predation. Some of them counteract conspicuousness by becoming unpalatable or poisonous with alkaloids derived from the plants they eat while still caterpillars, as discussed in Chapter 3. Other butterflies, not so protected, gain a certain immunity to predation by mimicking distasteful species. However, I believe that the wings themselves, apart from chemical defenses and warning coloration, make butterflies unattractive to birds, who seem to dislike those broad wings flapping in their faces while they laboriously remove them. When smaller, more readily swallowed insects are available, birds tend to avoid butterflies. The nicks and gaps often noticed in butterflies' wings and commonly attributed to birds may more often be caused by mice, shrews, amphibians, reptiles, or other insects, probably frequently at night when both butterflies and most birds are inactive, or when butterflies have just emerged from the chrysalis and cannot yet fly. These mutilations suggest another advantage in having such ample wings: they divert attacks from the vulnerable body to lifeless tissues; even with the loss of much wing surface, butterflies can fly. Although one can point to certain compensations for the hazards of wearing such large, conspicuous wings, they do not seem adequate to account for them. What could have promoted the evolution of butterflies' colorful wings? Probably the answer to this question is to be found in their mating habits.

Neither birds nor butterflies bear their bright colors directly upon their bodies or wings, but in lifeless outgrowths from them, the feathers of birds and the scales of butterflies, which are readily rubbed off. Although the plumage of birds keeps them warm and streamlines their bodies, the scales that cover butterflies' wings have no obvious use except to bear their colors. These colors are due in part to pigments and in part to the optical properties of the scales' fine structure. White, red, orange, and yellow are due to pigments; blue, as on the morpho's wings, is caused wholly by optical interference in thin films in the scales' outer layers. Green, at least at times, results from the combined effects of optical blue and yellow pigment. Among the pigments in butterflies' scales are waste products of metabolism. The white in the wings of the Cabbage Butterfly *(Pieris rapae)* and related species is uric acid, which instead of being excreted is stored in the lifeless scales. Yellow and orange are sometimes caused by lepidotic acid, a derivative of uric acid. To transmute waste products into beautiful colors is certainly an admirable economy.

We could proceed to study sexual selection in birds without first paying attention to their vision because, being vertebrates like ourselves, with simple eyes equipped with lenses, they see much as we do, but often better. We cannot profitably discuss mate selection in butterflies, so much more different from ourselves in many ways, without first asking: What do they see, and how well? When we read that butterflies have the widest visible spectrum known in the animal kingdom, ranging from ultraviolet to red (bees respond to ultraviolet but not to red or orange), and that they are especially sensitive to the basic wing coloration of their own species, we are inclined to believe that we have found the secret of their colorful, often intricately patterned wings.

However, when we pursue the subject further and learn the limitations of compound eyes, composed of many narrow light receptors called ommatidia, doubts arise. In general, for butterflies to distinguish two points as separate, they must be no closer together than several degrees of arc, which is hundreds of times coarser than the optical resolution of human eyes, about half a minute of arc. To appreciate what this means in terms of distinguishing two-dimensional patterns, we must square the value for simple angular resolution, which leads to the conclusion that for distinguishing details, the vision of butterflies may be tens of thousand of times less efficient than that of humans and other vertebrates. It follows that butterflies are shortsighted; they can clearly distinguish fine patterns and shapes only when very close. However, their sensitivity to movement is more acute than ours; they are well able to detect moving objects that differ in color, and especially in brightness, from the surrounding field. Moreover, we must keep in mind the fact that butterflies' sensitivity to ultraviolet radiation, which is reflected from the wings of many species, makes these wings appear to them otherwise than to us. Species whose color patterns appear similar to us may present quite different patterns to eyes sensitive to ultraviolet.

Among birds, females seek males who remain on their territories or stations in courtship assemblies, advertising their presence by voice, often accompanied by visual displays. Some male butterflies also assemble in leks, but among lepidopterans males are usually the active seekers. Female moths draw males to themselves by scents diffused through the night air. Diurnal male butterflies actively seek the opposite sex. Some search on wing for females lurking amid vegetation. Others wait in exposed situations and watch for passing females, or an individual male may change his procedure as occasion demands.

Male butterflies reveal the inadequacy of their eyesight by pursuing a wide range of inappropriate objects that evidently they mistake for females. F. A. Urquhart saw Monarchs *(Danaus plexippus)* follow seven other species of large butterflies, as well as a Chipping Sparrow *(Spizella passerina)*, a Song Sparrow *(Melospiza melodia)*, a leaf, and a scrap of paper blown by a dust devil. Yellowing leaves, fragments of white paper, even small white flowers incite a male Cabbage Butterfly to hover and investigate. A thorough study of a male butterfly's sexual pursuit was carried out with the Grayling *(Eumenis semele)* in the Netherlands by Niko Tinbergen and his colleagues. This protectively colored gray butterfly — "bark with wings," they called it — was seen to follow twenty-five species of lepidopterans, of the most diverse colors, plus such sundry insects of other orders as wasps, dragonflies, grasshoppers, and dung beetles, even birds from Chaffinches *(Fringilla coelebs)* to Mistle Thrushes *(Turdus viscivorus)*. In some fifty thousand tests with models dangled from the end of a rod, these researchers learned that the male Grayling will pursue pieces of paper of the most varied sizes, shapes, and colors from red to blue. The darker papers tended to release most responses; black was the most effective model, white the least. Models much larger than the butterfly were followed more frequently than those of the butterfly's size; like birds, butterflies respond strongly to supernormal stimuli. A fluttering movement, simulating a butterfly's flight, made the models more attractive.

Although a male butterfly's initial response to a moving or even stationary object that might be a female of his kind is undiscriminating, when his pursuit brings him near, the myopic insect can tell, apparently by color, whether the thing he has chased is a female of his species or something else. If he has overtaken a receptive female, the two join in a nuptial flight to a secluded spot amid vegetation, suitable for mating. Perhaps now, at close range, the female recognizes the male by his color pattern, which in many species is brighter or bolder than her own. Experimenters have tackled this problem, generally with negative results.

Thus, Robert E. Silberglied, in a carefully controlled experiment, blackened the wings of a number of individuals of the brilliant red butterfly *Anartia amathea* of Panama, and confined them in a flight cage. Of the twenty-one matings that resulted, eight were by red-winged females with untreated red-winged males, seven were by red-winged females with black-winged males. Five matings were by red-winged females with red-winged males whose wings had been coated with colorless varnish as an additional control. Only one red-winged male mated with a black-winged female. Evidently, the females were not influenced by the males' color, but the males paid attention to color and mostly avoided females whose red wings had been artificially blackened. Although the number of individuals in this experiment was small, the results are in accord with many other observations on the responses of female butterflies to males. Their acceptance of a male is not determined by visual clues, unless, perhaps, these are in the ultraviolet, which we cannot see.

Scent, rather than sight, appears to determine a female butterfly's response to a male, and perhaps, ultimately, his continuation of the mating ceremony to its consummation. In various positions on the forewings or hindwings of male butterflies are patches of scales, narrower than those elsewhere on the wings and sometimes terminated by tiny brushes, called androconia. Associated with glands, these scales bear the scents that males use to assure females of their correct identity and win acceptance. On some butterflies the scent glands are situated on the body rather than the wings. On the Monarch and related species, they are at the tip of the abdomen, and there is also a cuplike scent receptor on the upper side of a rear wing. The scents of male butterflies, flowerlike or otherwise pleasant, are often quite evident even to the relatively insensitive human nose.

When a male Grayling overtakes a flying female of his kind, he follows her in a wild pursuit. Often she escapes him, but sometimes both alight close together. After an elaborate courtship display, in which the male strikes the female with his wings, he bows beside her and firmly presses her antennae between his wings, bringing the sensitive knobs at their ends into contact with the scent organ on his left forewing. If she is virgin and remains, coition follows. Some butterflies stay clasped together for hours, during which the male inserts his sperm capsule into the female. After they separate, each goes its own way. The male butterfly's only contribution to reproduction is his sperm, and perhaps a little nourishment from the capsule, or spermatophore, in which it is enclosed.

Because female butterflies seem so little affected by the males' coloration, Silberglied, after carefully sifting a great mass of observational and experimental data, proposed the hypothesis that intrasexual communication between males, rather than intersexual or interspecific communication, is the major selective agent responsible not only for the brilliance of male butterflies' coloration but likewise for its constancy, relative to that of the often more variable females of the same species. Although in many species females but not males have different color phases, the reverse situation, variable males and constant females, does not occur. Likewise, in many species, females exhibit protective mimicry while males do not, and again the reverse of this situation is absent.

Male butterflies sometimes fight among themselves. A few have wings modified as weapons and may injure each other. It is to their advantage to avoid encounters, in which the winner as well as the loser may be seriously damaged. Instead of physical contact, they may engage in "psychological warfare," employing their colors as weapons. Probably more frequently, they tend to avoid one another, while they increase their chances of meeting responsive females by scattering over fields and woodland. Far from attracting each other, they are repelled by visual stimuli resembling other males, especially by white, ultraviolet reflectance, and iridescent colors. While mating, male Orange Sulphur Butterflies *(Colias eurytheme)* warn approaching intruders by flashing their wings.

In support of his hypothesis that intrasexual rather than intersexual selection is primarily responsible for the vivid coloration of male butterflies, Silberglied turned to writers who contend that the bright plumages of birds have evolved "strictly for aggressive signaling." Earlier chapters of this book have presented abundant evidence that female choice, rather than competition among males, is mainly responsible for the colorful plumage of the latter, especially its finer details. In any case, the situation among birds is very different from that among butterflies, for in the former females seek males, while in the latter males more often seek females. And even among butterflies, females are not invariably indifferent to the appearance of their suitors. In the same volume that contains Silberglied's excellent review of sexual selection in butterflies, David Smith presents painstakingly gathered evidence that in the African Monarch *(Danaus chrysippus,* a close relative of the North American Monarch), "female choice between morphs certainly occurs and the evidence points to its being actuated by visual rather than olfactory cues." However, he recognizes that both male competition and female choice operate in sexual selection. The question is not whether one or the other is exclusively responsible for secondary sexual characters, especially bright colors and lavish adornments, but which is the predominant factor. The relative importance of these two influences may vary from animal to animal.

The hypothesis that the brilliant coloration of male butterflies evolved primarily to prevent their encroachment upon each other's domains appears at least partly tenable to a naturalist who has enjoyed but never carefully studied these insects. It seems adequate to account for such large expanses of color as are exposed by morphos, certain birdwings, and Orange Sulphurs. But a hypothesis that regards the coloration of butterflies as serving primarily to promote avoidance rather than close approach seems inadequate to explain the intricate patterns, the fine details, so frequent in butterflies' wings. Compound eyes with poor resolving power would seem adequate to distinguish these details, if at all, only when very near. Other agents of selection have helped to paint butterflies' wings. The eyespots so frequent near the wings' margins serve to deflect predatory attacks from the vulnerable body. Fine mottling helps some species to blend with the surface on which they rest. Bold stripes may break the outline as seen at a distance. A rhythmic deposition of pigments may create patterns that are not adaptive. Doubtless features that make butterflies attractive to us are lost on the insects themselves; just as others that attract butterflies to sexual partners are indistinguishable by our eyes with a narrower visual spectrum. If we were as sensitive to ultraviolet light and to scents as butterflies appear to be, much that remains enigmatic about their courtship should become clearer.

References

Newbigin 1898; Richards 1927; Silberglied 1984; Smith 1984; Tinbergen 1951; 1958; Urquhart 1960; Vane-Wright and Ackery 1984.

18.
Do Birds Choose Beauty?

In the preceding chapters we reviewed the courtship of butterflies and, especially, birds, as reported with scientific objectivity by many investigators. Now we turn to aspects of mate selection not amenable to strictly scientific investigation because they involve the psychic life of animals, which cannot be observed either directly or by any instrument presently available, but which is not for that reason of less interest or importance than the objective phenomena. We must ask questions such as "Are animals capable of true choice?" "Have they aesthetic sensibility?" "Do they enjoy beauty?" As hitherto, we shall give attention chiefly to birds, for which we have the most abundant data, and which, like ourselves, have keen vision and (with some outstanding exceptions) a rather obtuse sense of smell, rather than the reverse, as in many mammals and invertebrates.

Scientists and philosophers alike have doubted the reality of sexual selection, as conceived by Darwin, because it implies a refinement of choice that they are reluctant to attribute to birds. Since this is the crux of the matter, let us tackle this question by first considering how we ourselves choose. I enter a shop to buy a shirt like those that I have long worn on the farm and in the forest. The attendant shows me several that do not match my mental image of what I want. I hardly even consider them. Finally, he lays upon the counter a shirt just like the one I wore yesterday. Without delay, I accept it. I have not really chosen; at most, I have made a rapid judgment, but my reaction to the shirt hardly deserves that designation. It was more like the reaction of an animal who responds immediately to an appropriate stimulus. To use the terminology of the ethologists, the shirt acted as a "releaser" (of some of my money!); my mental image of the style I desired corresponded to "an innate releasing mechanism." Allowing for slight differences, my behavior was comparable to that of a Herring Gull chick who begs when presented with a red spot more or less like the red mark near the end of its parent's lower mandible, or of a European Robin who becomes aggressive when confronted with a red patch like that on his own breast. A releaser corresponds to a key; the innate releasing mechanism to the lock which, if not rusty or otherwise out of order, always yields to a key of the proper shape, and to no other. Like a rusty lock, an animal not in the appropriate physiological state does not respond to a releaser; when satiated, it fails to react to food; when not sexually receptive, it is unmoved by a courtship display.

The storekeeper from whom I have bought the shirt has just received new stock and persuades me to buy another shirt for "dress." He spreads before me several with patterns that I like. It takes me a while to decide which I prefer. Finally, I select one; but in the absence of this particular shirt, I would have taken another. I have made a true choice. To generalize: if an animal presented with A, B, and C selects A at a time when it would not have accepted B or C, it can hardly be said to have chosen. But if A, B, and C are simultaneously offered and it consistently chooses A, although in the absence of A it would have accepted B or C, and especially if it pauses before deciding, it has made a true choice. Choosing is more than accepting an item in the absence of an alternative.

Now we must ask whether the behavior of a female bird visiting a courtship assembly resembles mine when I bought the first shirt or when I bought the second. Undoubtedly, she has a mental image — whether highly or dimly conscious we need not decide — of the male of her species; only in exceptional circumstances would she mate with one of a different species, leading to hybridization. If the assembly contains only one male who conforms to her "specifications," she does not choose so much as seek him out; she judges his acceptability. But if several males who meet her requirements are present, so that in the absence of the one she mates with she would have taken another, less attractive but adequate, she has made a true choice.

A female about to lay her egg(s) sometimes, perhaps regularly, visits an assembly for a preliminary inspection of its members a day or more before she is ready to mate. When ready for coition, she may walk or fly past several eligible males to the one she prefers, who is frequently at or near the center of the assemblage. Her preference for a centrally situated male is probably genetically determined, because as males increase in age, experience, and vigor, they work their way inward. But several males of approximately equal status may occupy the center, and she must choose between them. If for any reason they do not attend her, she may accept one lower in the hierarchy as she leaves, as has been observed in the Sage Grouse. Above all, among Ruffs, who not only differ strikingly in appearance but in the presence or absence of a satellite, position in the lek, and other details, multiple factors influence the reeves' decisions. Even when we can detect no difference between adult members of an assembly, as in small birds like manakins, females probably notice differences; it is well established that birds recognize individuals in flocks of birds indistinguishable even by experienced ornithologists. Ducks cannot be compelled to accept a partner whom they do not prefer. Clearly, the first eligible male that a female encounters does not release the mating reaction by tripping an innate releasing mechanism. Her choice is determined by a constellation of factors that influence her mind, including the appearance and displays of the male, and perhaps other features more subtle.

Those who question whether birds are capable of making true choices are frequently reluctant to concede that they have an aesthetic sense. We cannot ask birds whether they enjoy bright colors or melodious sounds, but perhaps we can resolve our doubts without awaiting a reply. We often decide whether humans have aesthetic sensibility by purely objective criteria, including the following: Do they prefer a beautiful or handsome spouse? Do they prefer decorated artifacts to unadorned ones that would serve them equally well? Do they embellish their homes and surroundings? Do they create beauty, as by painting pictures, composing music, planting a flower garden, embroidering a fabric, or the like? Do they sing, play musical instruments, or frequently listen to music? If the answer to any of these questions is "yes," we conclude that a person is not devoid of aesthetic sensibility.

Let us apply these criteria to birds. Darwin attributed the splendor of male birds to the preferences of females, and many subsequent naturalists have concurred. Others disagree, asserting that male birds, especially those in courtship assemblies, have developed their bright colors and adornments to impress and dominate competing males. We looked briefly at this explanation in Chapter 14. That brilliant attire plays a part in intrasexual confrontations I do not deny, but I have seen no convincing evidence that this is the principal role of their adornments. In any case, to ascribe the evolution of male adornments to intrasexual rather than intersexual selection appears to make similar assumptions about the aesthetic sensibility of birds, for now the male bird's beauty is held to impress other males more powerfully than it impresses females. A male reveals the aesthetic effect of an outstandingly handsome individual of his own sex by recognizing his superiority and accepting a subordinate role; a female, by mating with him. However, no matter how dominant a male bird may be, he cannot compel a female to accept him; therefore, we cannot escape the conclusion that her preference for the male who wears the more beautiful plumage, or displays it more charmingly, is the proximate factor in the evolution of the male's adornments. Our first criterion for the presence of an aesthetic sense in birds is satisfied.

Crows, magpies, and jays reveal a nascent aesthetic sense by occasionally carrying off and hiding shining artifacts of metal, such as coins, keys, and needles, as well as buttons, shells, scraps of colored cloth, or anything lustrous or colorful they can find and hold in their strong bills — a habit associated with their custom of making caches of excess food to which they return in times of need. Bowerbirds, which like birds of paradise are held to be rather closely related to the corvids, carry this tendency much farther, regularly decorating their constructions with a variety of colorful or shining objects — feathers, fruits, shells, bleached bones, small human artifacts. Moreover, some of the avenue builders paint their walls. However, by spreading their decorations more or less at random over their platforms, they reveal that their taste is not very refined. A few of the gardener bowerbirds show superior taste by arranging fruits, flowers, or fungi in neat piles, each composed of objects of the same color, and removing them as they wither or decay. By building such a charming pavilion, fronting such a pretty garden, the Vogelkop Gardener creates beauty, no less than humans who plant flowers in front of their houses. Bowerbirds meet our second and third criteria of an aesthetic sense.

Birds might be said to create beauty indirectly through the evolutionary process, by selecting beautiful sexual partners. This major contribution to the beauty of the natural world is due chiefly to females. Males create beauty directly by their singing. The songs of birds, so appealing and varied, provide our most compelling reasons for believing that birds have a sense of beauty, and for a rather obvious consideration. Lacking mirrors, birds cannot see much of themselves, and it may not occur to them that they look like others of their species and sex. No individual bird can do much to change its appearance except by molting, a physiological process as little under voluntary control as the growth of our hair. The best a bird can do is to keep its plumage clean and tidy, by frequently preening and anointing it with oil from its preen gland. But a bird can hear itself singing; it can sing voluntarily; and it can change its tune to suit its mood or taste. In terms of value, a singing bird is simultaneously a value generator and a value enjoyer. By singing, it can generate value both for itself and for those who hear and enjoy its notes. Song is the most widespread way that birds directly or voluntarily create aesthetic value. Other ways include the building and adornment of bowers, as is done by very few birds, mostly in remote forests, and the decoration of charming nests with lichens or moss, as many birds do, more probably for camouflage than for aesthetic reasons.

Birds appear to enjoy not only singing but also listening to the notes of other birds and even nonvocal sounds. Why should mimics take such pains to learn and imitate sounds if they took no pleasure in them? Mimicry has much in common with play. Animals play by performing, for immediate enjoyment rather than utilitarian ends, activities at which they are adept. Wide-winged birds soar on rising air currents; well-fed, rested horses frolic by galloping over the pasture; dolphins line up and race just ahead of an advancing ship. Birds with versatile voices may sing as other animals play, for the pleasure of engaging in an activity at which they excel, with the sound of their own voices as an added reward.

I have little doubt that birds sing because they enjoy hearing themselves. During four months I lived in a plantation house on a hilltop, surrounded by coconut palms amid whose wide-spreading fronds a numerous company of Great-tailed Grackles (*Quiscalis mexicanus*) roosted and nested. Although the yellow-eyed males, in sleek black plumage glossed with violet and blue, lacked a distinctive song, the range and power of their voices was impressive. At one extreme, they rapidly repeated little tinkling notes; at the other, they called so loudly and piercingly that they were best heard at a distance. In addition to commonplace phrases, from time to time I noticed a new verse, attractively simple or buglelike and stirring, which was evidently an original invention of one of the males. For a week or so, I would hear this innovation over and over, until, like a popular human song, it grew stale and was forgotten. These promiscuous grackles did not defend territories, nor were they courting when they repeated their notes; they seemed to exercise their voices only for their own amusement. Similarly, males of a related nonterritorial, nonpairing colonial nester, the Yellow-rumped Cacique *(Cacicus cela),* resting idly among the females' swinging pouches, repeat a variety of delightfully bright short phrases, for no apparent reason except their own delectation. Likewise, mimics such as the Northern Mockingbird, Blue-and-White Mockingbird, and Black Thrush seem to delight in the display of their virtuosity. Even domestic hens, well fed and content, sing with cheerful disregard of melody.

In former times, people believed that birds sang because they were happy. In Plato's dialogue, the *Phaedo,* Socrates declared that "no bird sings when it is hungry, or cold, or in any pain; not even the nightingale, nor the swallow, nor the hoopoe, which, they assert, wail and sing for grief." To be sure, birds sing in cages when they have little reason to be happy, and they have been cruelly blinded to make them more songful. A Guatemalan girl offered a succinct explanation of this paradox: *"Can-tan para no llorar"* (They sing to avoid weeping). Deprived of freedom, even of sight, captive and mutilated birds bravely comfort themselves with the only resource left to them: they sing, as in Thomas Hardy's moving poem, "The Blind Bird."

The ancient notion that bird song is primarily an emotional outpouring, and above all a pure expression of happiness or contentment, has needed revision in the light of modern studies. When Bernard Altum established that bird song serves the important utilitarian functions of proclaiming possession of a territory and attracting a mate, he too hastily concluded that birds never sing for their own enjoyment. To assert that an organ or an activity cannot serve multiple purposes is as absurd as to declare that because we use our hands to feed ourselves we cannot write with them. The very activities most indispensable to an animal are precisely those most often used for enjoyment in intervals of leisure, as when birds soar, horses run, and people exercise their minds solving puzzles or playing chess. Moreover, the utility of birds' songs fails to account for the musical excellence of many of them. A series of harsh or raucous notes, if sufficiently loud and distinctive, might serve equally well to proclaim territory and attract a mate, if the latter were devoid of aesthetic sensibility and did not prefer something better than mere noise. Indeed, an experienced ornithologist can point to birds whose "songs" are only technically so called yet serve the biological needs of their species. We need not suppose that the sense of beauty is equally developed in all birds, as it certainly is not in all humans. Songful birds, who not infrequently invent new verses, satisfy our third and fourth criteria for aesthetic sensibility.

No one has presented better reasons for believing that birds have aesthetic feeling, that the songs of many are music, and that they enjoy singing than Charles Hartshorne in *Born to Sing* and other writings. He was, I believe, the first to recognize the "monotony threshold," which he also called, more clearly, the "anti-monotony principle." The songs of some birds are so varied that, like the performance of an orchestra, they flow on and on without becoming tiresome. Other birds have very limited repertoires, perhaps only a single verse, which they repeat over and over. To avoid monotony, such birds tend to intercalate intervals of silence between their verses, so that each repetition will fall upon their ears, and those of their hearers, with a certain freshness. Although of wide application, the anti-monotony principle is not universally valid. Surprisingly, a bird that violates it conspicuously is a member of a family renowned for song. For reasons unexplained, the Mountain Thrush (*Turdus plebejus*) of Middle America is a very inferior singer, who with scarcely a pause pours out a seemingly interminable succession of weak, lusterless notes, with little variation in pitch and the merest suggestion of rhythm: *chip chip chip cher chip chip cher chip chip chip* over and over. The dawn songs of certain American flycatchers are also monotonous. At daybreak in the highlands of Middle America, the little Yellowish Flycatcher (*Empidonax flavescens*) repeats *seee seee chit*, the first two notes weak and sibilant, the third sharper, at the rate of about twenty-one times per minute, with scarcely a pause between repetitions. This uninspired recital often continues for a quarter of an hour or more, until dawn's increasing brightness silences it, usually for the rest of the day.

When we survey all the evidence — elegant plumage due to female choice, tastefully decorated bowers, melodious songs, the creation of beauty as well as attraction to it — support for the thesis that birds have aesthetic sensibility becomes irresistibly strong. One who rejects this conclusion must resort to an ad hoc hypothesis for each category of beauty, a philosophically unsatisfactory procedure. The capacity to have life quickened by colors, forms, sounds, and movements, most strongly when they are most beautiful, harmonious, or rhythmic, appears to be the neurophysiological foundation of the aesthetic sense. About the feelings of birds or other creatures when excited by beauty, we can only conjecture, but it would be strange if they were not pleasant.

The widespread reluctance to admit that birds have an aesthetic sense, and that this influences their selection of mates, is but a special case of a much more widespread phenomenon, human materialization of the nonhuman world, our blindness to its psychic aspect. Primitive people did not doubt that the animals around them, even trees, rivers, and mountains, had souls and feelings, or were the abodes of invisible spirits. They apologized to the animal they slew, alleging their great need of its flesh; they prepared a habitation for the spirit of the tree they were about to fell; they made a votive offering to the spring from which they drew water. Their animism was a groping expression of a profound intuition, a more penetrating insight than our modern materialism. But as, with growing technological competence and urbanization, humans receded spiritually ever farther from nature while making ever greater demands upon its productivity, their attitude toward the creatures around them hardened. Their interest centered ever more narrowly upon the material aspects of everything, animate or inanimate, that they exploited for their needs or pleasure. Not the mind or feelings of the slaughtered animal but its flesh or other useful products, not the dryad who might dwell in a tree but its wood, interested them. Technological humans materialize what they exploit.

It is a curious fact that while human understanding of natural processes, and human ability to control and reap benefit from them, grew apace, humankind withdrew spiritually farther from the rest of the living world — with the exception of a few perceptive individuals outside the mainstream. It is highly significant that the most influential philosophical expression of human aloofness from nature, Descartes' foolish doctrine of animal automatism that his younger contemporary, Spinoza, rejected, was published early in the modern era that has seen such great advances in science and technology.

Modern human materialization of nature has created a prejudice against the ascription of higher psychic attributes, in whatever degree, to nonhuman animals. It has made us underrate the mental capacities of birds and mammals, some of whom give clearer indications of intelligence the more intimately we know them. Too often, inappropriate methods have caused experimenters to underestimate the minds of their subjects. When we treat animals with more understanding, they frequently reveal mental capacities that surprise us. An excellent example is Irene Pepperberg's study of vocal learning by the African Gray Parrot (*Psittacus erithacus*). Parrots are commonly believed to repeat human words without understanding; to "talk like a parrot" is to chatter senselessly. But by employing social modeling — using live, interacting human tutors — Pepperberg taught her parrot to request, refuse, identify, categorize, or quantify more than sixty items, and to use with understanding phrases such as "come here," "I want x," and "wanna go y."

We should be at least as cautious in setting arbitrary limits to the mental powers of animals as in making unsubstantiated claims for them. One who reflects that small birds can navigate twice a year, by means that we only partly understand, between definite points thousands of miles apart, will hesitate to depreciate their mentality. Fortunately, the old prejudices are weakening as an increasing number of well-prepared people give more serious attention to animals whose habits have not been distorted by long domestication. The more we study the courtship of birds, the more convinced we become that Darwin was right when he attributed the adornments of birds and other animals to their choice of mates, which implies that they are not devoid of aesthetic sensibility.

References

Hartshorne 1973; Mayr 1935 (for Altum); Pepperberg 1985.

Epilogue:
Quality in a Sea of Quantity

This book addresses a neglected question: How can a process dominated by quantity yield quality that is more than adaptation, however finely attuned, to an environment and a style of living? How can evolution, which produces such a vast diversity of creatures regardless of high values, yield such values? How can excellence arise and persist in a world swarming with organisms whose only merit appears to be their ability to survive and multiply, at whatever cost to surrounding creatures? This is the problem which the foregoing chapters have attempted to answer.

High values are aesthetic, moral, and intellectual, in Western culture traditionally known as the beautiful, the good, and the true. Life and health, precious though they are, are basic goods rather than high values; their worth depends upon how they are used. Beauty and goodness, which I take to mean the ability to live in harmony with surrounding beings, are widespread in the living world; intellectual value, the quest of knowledge and understanding is, largely if not wholly, confined to humanity. The present book is limited to an investigation of the origins of beauty, a subject wide and complex enough to fill it.

A major conclusion of our survey of the origins of natural beauty is that we owe a great part of it to cooperation and mutually beneficial interactions among organisms. Flowers became colorful and fragrant to attract pollinators that they rewarded with nectar and nutritious excess pollen. Fruits developed color and aroma to advertise their availability to birds and other animals who disseminate their seeds. The splendor of birds has been enhanced by a very different mode of cooperation, that of males in courtship assemblies and other mating arrangements that enable females to choose freely. To be sure, while males cooperate to attract females to a well-known mating center, they compete keenly for the privilege of inseminating these females, and subordinate members of the gathering may, in a given year, be virtually excluded from mating. But, on the whole, this method of courtship benefits all participants, for as they grow older, subordinates may rise to higher status. At least, if they escape predation, disease, and other sources of mortality, they survive from season to season, as might not be true if the ritualized conflicts between males had not supplanted crude and lethal fighting.

Although the simultaneous action of cooperation and competition has not been widely recognized, it is frequent in nature as in human society, and, on the whole, beneficial. In tropical rain forest, trees cooperate to create the environment indispensable for the growth of most of them, while they compete silently for space in the sunlit canopy. In a civilized community, responsible citizens cooperate to keep it orderly and healthful, while they compete for economic advantages. A contest between athletic teams degenerates into a melee if the players do not cooperate to observe the rules while they compete for victory.

In the foregoing modes of peaceful cooperation, we recognize moral value, for morality is, above all, the endeavor to bring harmony into life. To attribute moral value to the flower–pollinator–seed–disseminator cycle, or to a pacific courtship assembly of birds, is not tantamount to declaring that all or any of the creatures involved in them have a developed morality; it is merely to recognize that these are associations which a wide-eyed morality must approve. Although no nonhuman animal appears to have a self-conscious, foreseeing morality attentive to maxims, in some of them we recognize what I have called protomorality, the germs from which our ethical concern has grown. Neither aesthetic sensitivity nor morality sprang into the world fully formed, like Athene from the head of Zeus.

This brings us to the second of the high values, goodness, which is harmony in every aspect of life. In an earlier book, I dealt with one of the higher developments in this sphere among vertebrate animals, the cooperative breeding associations widespread among birds, especially in the tropics and subtropics where they reside permanently. In these associations, progeny remain with their parents for one or more years, helping them to rear later broods. The nestlings or chicks whom the helpers attend are usually their younger brothers and sisters, but not always, for frequently a bird of other parentage joins the cooperating group. Members of these associations live together in harmony, often preening one another, exchanging food, sleeping in contact on a branch or in a dormitory nest that they build. All grown members of the group join in defending their territory from encroachments by neighboring groups, nearly always by formal displays that avoid crude fighting. Co-operative breeding is the highest expression of avian society.

The widespread trend among the vertebrates to produce fewer young and take better care of them has favored the emergence of quality with reduction of quantity. Care of dependent offspring, whether by coopera-tive groups, male and female pairs, or single parents, has been the seed-bed of moral virtues. In devoted parental care, by whatever animal, we detect glimmerings of responsibility, sense of duty, generosity, sympathy, and love. When both parents cooperate closely in attending their off-spring, bonds not unlike those which bind them to their young may grow up between them. This is the context in which love between the sexes that is more than a transient passion was born. When we remember all that caring for progeny has done to mellow and ennoble the human spirit, we reflect with a pang that overwrought sexuality and excessive reproduction, by throwing animals into savage competition for inade-quate resources, has been a major source of the corrosive passions, the violence, and the ugliness that afflict life. If evolution, instead of becom-ing dominated by quantity, had been more sensitive to quality and de-veloped widespread restraints upon reproduction, as by feedbacks that adjusted population to resources, it might have made the living world more peaceful and beautiful.

In this book we set out to find a partial answer to the question: How can a process dominated by quantity yield high quality? How can aes-thetic and moral values emerge from a process indifferent to values? We have recognized and described certain modes of cooperation among or-ganisms that help them to become beautiful. Although we have tried to account for these developments by widely accepted evolutionary prin-ciples, essentially mutation and selection, something more appears to be needed fully to explain the results. Mutations change the genetic consti-tutions of organisms and through these alterations their structures and functions; selection removes the unfit while sparing individuals well adapted to their habitats. Neither is a true constructive principle. How can mutation and selection promote aesthetic sensibility, love, sympathy, and similar affections?

The answer to our dilemma is that organic evolution is not an inde-pendent or self-sustaining process but a complication that has arisen in the cosmic process of harmonization, which builds up the components of the Universe into patterns of increasing amplitude, complexity, and coherence. In the living world it operates primarily as growth, which unites molecules from soil, water, and air in tightly integrated organisms capable of a diversity of functions. Without growth, the true creative

Pheasant-tailed Jacana, *Hydrophasianus chirurgus*. Sexes similar. India and Malaya to Java and the Philippines.

principle in evolution, mutation would accomplish nothing and natural selection would lack creatures to select. Harmonization is the powerful ongoing stream that raises Being to higher levels of organization and value; evolution diversifies this stream. It might be compared to a prism that spreads a beam of light into a spectrum of many colors, to an obstacle that shatters a strong jet of water into a multitude of drops flying in all directions. Evolution is centrifugal, harmonization centripetal. Without harmonization's strong tendency to integrate, to unify, to bring concord into the living world, evolution might yield only unmitigated discord. Evolution gives life immense variety, harmonization impels it onward and upward.

It is not difficult to understand how harmonization produces high values because such values are modes of harmony. This is most evident in music, in which melody and aesthetic value are produced by harmony of sounds. Visual beauty depends upon the harmonious arrangement of forms and colors, which should contrast without clashing. Moreover, nothing can be truly beautiful unless it harmonizes with the rhythms, the innate preferences, or the acquired taste of a receptive mind. Goodness, or moral value, is a harmonious relationship with the living things that surround us. Intellectual values, largely confined to humanity, are generated by our quest of knowledge or truth. Our most trustworthy criterion of truth is the coherence of ideas. When the testimony of our senses contradicts our beliefs, when supposed fact is irreconcilable with supposed fact, when theory conflicts with theory, we are confused and uncertain; when all our observations and interpretations, at least about some small segment of reality, are compatible and form a coherent body of thought, we believe that we have found truth and enjoy the high value of intellectual clarity. As beauty depends upon harmony among sensuous impressions, goodness upon harmonious relations with surrounding beings, so truth, for us who lack absolute knowledge, is harmony among the contents of our minds.

We do not hesitate to express aesthetic judgments about nature, but there is a widespread feeling that to judge it by ethical criteria is inappropriate. However, if we bear constantly in mind that the essence of moral goodness is harmony among living beings, our moral judgments can be more objective than our aesthetic judgments, which express personal preferences, so that what one person calls beautiful may appear plain or ugly to another. As parts of nature, products of the same evolutionary forces that have shaped all its other parts, we have an inalienable right to judge it by our highest ethical standards. To refuse to assess it so is to shackle one of evolution's most precious achievements, our sense of right and wrong, which impels us to condemn the crudities for which evolution is responsible. When, liberating our minds from the mistaken notion that moral judgments about nature are less admissible than aesthetic judgments, we view it with understanding, we find that a large part of its beauty has been promoted by mutually beneficial relations between organisms, of the same or different species, that are morally admirable. Of this beneficent association of aesthetic and moral values, of the beautiful and the good, a lovely flower is an appropriate symbol.

Bibliography

Armbruster, W. S., and G. L. Webster. 1979. Pollination of two species of *Dalechampia* (Euphorbiaceae) in Mexico by euglossine bees. *Biotropica* 11:278–283.

Armstrong, E. A. 1963. *A study of bird song.* London: Oxford University Press.

Bateson, P., ed. 1983. *Mate choice.* Cambridge, England: Cambridge University Press.

Beehler, B. 1983a. Frugivory and polygamy in birds of paradise. *Auk* 100:1–12.

———. 1983b. Lek behavior of the Lesser Bird of Paradise. *Auk* 100:992–995.

———. 1987a. Ecology and behavior of the Buff-tailed Sicklebill (Paradisaeidae: *Epimachus albertisi*). *Auk* 104:48–55.

———. 1987b. Birds of paradise and mating system theory — predictions and observations. *Emu* 87:78–89.

———. 1988. Lek behavior of the Raggiana Bird of Paradise. *Natl. Geographic Research* 4:343–358.

Beehler, B. M., and M. S. Foster. 1988. Hotshots, hotspots, and female preference in the organization of lek mating systems. *Amer. Naturalist* 131:203–219.

Beehler, B., and S. G. Pruett-Jones. 1983. Display dispersion and diet of birds of paradise: A comparison of nine species. *Behav. Ecol. Sociobiol.* 13:229–238.

Belt, T. 1888. *A naturalist in Nicaragua.* 2d ed. London: Edward Bumpus.

Bennett, L. J. 1938. *The Blue-winged Teal: Its ecology and management.* Ames, Iowa: Collegiate Press.

Bent, A. C. 1923. Life histories of North American wildfowl, Part I. *U.S. Natl. Mus. Bull.* 126.

———. 1925. Life histories of North American wildfowl, Part II. *U.S. Natl. Mus. Bull.* 130.

———. 1926. Life histories of North American marsh birds. *U.S. Natl. Mus. Bull.* 135.

———. 1932. Life histories of North American gallinaceous birds. *U.S. Natl. Mus. Bull.* 162.

————. 1940. Life histories of North American cuckoos, goatsuckers, humming-birds and their allies. *U.S. Natl. Mus. Bull.* 176.

————. 1942. Life histories of North American flycatchers, larks, swallows, and their allies. *U.S. Natl. Mus. Bull.* 179.

————. 1950. Life histories of North American wagtails, shrikes, vireos, and their allies. *U.S. Natl. Mus. Bull.* 197.

————. 1953. Life histories of North American wood warblers. *U.S. Natl. Mus. Bull.* 203.

Bent, A. C., et al. 1968. Life histories of North American cardinals, grosbeaks, buntings, towhees, finches, sparrows, and allies. Edited by O. L. Austin, Jr. 3 vols. *U.S. Natl. Mus. Bull.* 237.

Bluhm, C. K. 1985. Mate preferences and mating patterns in Canvasbacks (*Aythya valisineria*). In Gowaty and Mock, eds. 1985.

Borgia, G. 1985a. Bower destruction and sexual competition in the Satin Bowerbird (*Ptilonorhynchus violaceus*). *Behav. Ecol. Sociobiol.* 18:91–100.

————. 1985b. Bower quality, number of decorations and mating success of male Satin Bowerbirds (*Ptilonorhynchus violaceus*): An experimental analysis. *Anim. Behav.* 33:266–271.

Brower, L. P. 1969. Ecological chemistry. *Scient. Amer.* 220:23–30. Abstract in *Ibis* 112:138 (1970).

Burley, N. 1985. The organization of behavior and the evolution of sexually selected traits. In Gowaty and Mock, eds. 1985.

Campbell, B., and E. Lack, eds. 1985. *A dictionary of birds.* Calton, England: T. and A. D. Poyser.

Carpenter, F. L. 1976. Ecology and evolution of an Andean hummingbird (*Oreotrochilus estella*). *Univ. California Publ. Zool.* 106:1–75.

Catchpole, C. K. 1980. Sexual selection and the evolution of complex songs among European warblers of the genus *Acrocephalus*. *Behavior* 74:149–165.

————. 1985. Vocalization. In Campbell and Lack, eds. 1985.

Chapman, F. M. 1935. The courtship of Gould's Manakin (*Manacus vitellinus vitellinus*) on Barro Colorado Island, Canal Zone. *Bull. Amer. Mus. Nat. Hist.* 68:471–525.

Cínat-Tomson, H. 1927. Sur la sélection sexuelle chez la perruche (*Melopsittacus undulatus Schaw*). *Compt. Rend. Soc. Biol.* (Paris). 97:253–255.

Coward, T. A. 1928. *Birds of the British Isles and their eggs.* 3d ed. London: Frederick Warne and Co.

Darwin, C. 1871. *The descent of man and selection in relation to sex.* New York: Modern Library reprint.

Davis, T. A. W. 1949a. Display of White-throated Manakins *Corapipo gutturalis*. *Ibis* 91:146–147.

————. 1949b. Field notes on the Orange-crested Manakin *Neopelma chrysocephalum* (Pelz.). *Ibis* 91:345–350.

————. 1958. The displays and nests of three forest hummingbirds of British Guiana. *Ibis* 100:31–39.

Davis, T. H. 1982. A flight-song display of White-throated Manakins. *Wilson Bull.* 94:594–595.

Davison, G. W. H. 1981. Sexual selection and the mating system of *Argusianus argus* (Aves: Phasianidae). *Biol. Journ. Linn. Soc.* 15:91–104.

Diamond, J. 1986. Animal art: Variation in bower decorating style among male bowerbirds *Amblyornis inornatus. Proc. Natl. Acad. Sci. USA.* 83:3042–3046.

Dinsmore, J. J. 1970. Courtship behavior of the Greater Bird of Paradise. *Auk* 87:305–321.

Dorst, J. 1956. Etude biologique des trochilidés des hauts plateaux péruviens. *L'Oiseau et R.F.O.* 26:165–193.

———. 1962. Nouvelles recherches biologiques sur les trochilidés des hautes Andes péruviennes. *L'Oiseau et R.F.O.* 32:95–126.

Dowsett-Lemaire, F. 1979. The imitative range of the song of the Marsh Warbler *Acrocephalus palustris,* with special reference to imitations of African birds. *Ibis* 121:453–468.

Erskine, A. J. 1972. *Buffleheads.* Canadian Wildlife Series 4. Ottawa.

Finn, F. 1907. *Ornithological and other oddities.* London.

Forshaw, J. M., and W. T. Cooper. 1977. *Parrots of the world.* Neptune, N.J.: T.F.H. publications.

Foster, M. S. 1977. Odd couples in manakins: A study of social organization and cooperative breeding in *Chiroxiphia linearis. Amer. Naturalist* 111:845–853.

———. 1981. Cooperative breeding and social organization of the Swallow-tailed Manakin (*Chiroxiphia caudata*). *Behav. Ecol. Sociobiol.* 9:167–177.

Gaymer, R. 1982. *Two in a bush.* Oakville, Ontario: published by author.

Gibson, R. M., and J. W. Bradbury. 1987. Lek organization in Sage Grouse: Variations on a territorial theme. *Auk* 104:77–84.

Gilbert, L. E., and P. H. Raven, eds. 1975. *Coevolution of animals and plants.* Austin: University of Texas Press.

Gilliard, E. T. 1956. Bower ornamentation versus plumage characters in bowerbirds. *Auk* 73:450–451.

———. 1959a. Notes on the courtship behavior of the Blue-backed Manakin (*Chiroxiphia pareola*). *Amer. Mus. Novitates* 1942:1–19.

———. 1959b. The courtship behavior of Sanford's Bowerbird (*Archboldia sanfordi*). *Amer. Mus. Novitates* 1935:1–18.

———. 1959c. A comparative analysis of courtship movements in closely allied bowerbirds of the genus *Chlamydera. Amer. Mus. Novitates* 1936:1–8.

———. 1969. *Birds of paradise and bower birds.* London: Weidenfeld and Nicolson.

Gowaty, P. A., and D. W. Mock, eds. 1985. Avian monogamy. *Amer. Ornith. Union, Ornith. Monogr.* No. 37:1–121.

Grant, K. A., and V. Grant. 1968. *Hummingbirds and their flowers.* New York: Columbia University Press.

Grey of Fallodon, Viscount. 1927. *The charm of birds.* New York: Frederick A. Stokes Co.

Hamilton, W. J., III. 1965. Sun-oriented display of the Anna's Hummingbird. *Wilson Bull.* 77:38–44.

Hardy, J. W. 1963. Epigamic and reproductive behavior of the Orange-fronted Parakeet. *Condor* 65:169–199.

Hartshorne, C. 1973. *Born to sing: An interpretation and world survey of bird song.* Bloomington: Indiana University Press.

Höhn, E. O. 1953. Display and mating of the Black Grouse *Lyurus tetrix* (L.). *British Journ. Anim. Behav.* 1:48–58.

Howard, R. D. 1974. The influence of sexual selection and interspecific competition on mockingbird song (*Mimus polyglottos*). *Evolution* 28:428–438.

Hudson, W. H. 1920. *Birds of La Plata.* 2 vols. London: J. M. Dent and Sons.

Huxley, J. S. 1923. *Essays of a biologist.* New York: Alfred A. Knopf.

———. 1938a. Darwin's theory of sexual selection and the data subsumed by it, in the light of recent research. *Amer. Naturalist* 72:416–433.

———. 1938b. The present standing of the theory of sexual selection. In *Evolution,* edited by G. R. de Beer. London: Oxford University Press.

Huxley, J. S., and F. A. Montague. 1926. Studies on the courtship and sexual life of birds, VI: The Black-tailed Godwit, *Limosa limosa. Ibis,* ser. 12, 2:1–25.

Immelmann, K. 1962. Beiträge zu einer vergleichenden Biologie australischer Prachtfinken (Spermestidae). *Zool. Jb. Syst.* 90:1–196.

Kamil, A. C., and C. van Riper III. 1982. Within-territory division of foraging space by male and female Amakihi (*Loxops virens*). *Condor* 84:117–119.

Kearton, C. 1961. *Penguin island.* Newton, Mass.: Charles T. Branford Co.

Lank, D. B., and C. M. Smith. 1987. Conditional lekking in Ruff (*Philomachus pugnax*). *Behav. Ecol. Sociobiol.* 20:137–145. Abstract in *Auk* 105 (4 Suppl. 5D, 1988).

LeCroy, M. 1981. The genus *Paradisaea: Display and evolution. Amer. Mus. Novitates* 2714:1–52.

LeCroy, M., A. Kulupi, and W. S. Peckover. 1980. Goldie's Bird of Paradise: Display, natural history, and traditional relationships of people to the bird. *Wilson Bull.* 92:289–301.

Lill, A. 1979. An assessment of male parental investment and pair bonding in the polygamous Superb Lyrebird. *Auk* 96:489–498.

———. 1985. Lyrebird. In Campbell and Lack, eds. 1985.

Loffredo, C. A., and G. Borgia. 1986. Male courtship vocalizations as cues for mate choice in the Satin Bowerbird (*Ptilonorhynchus violaceus*). *Auk* 103:189–195.

Low, J. B. 1945. Ecology and management of the Redhead, *Nyroca americana,* in Iowa. *Ecol. Monogr.* 15:35–69.

McKinney, F. 1985. Primary and secondary male reproductive strategies of dabbling ducks. In Gowaty and Mock, eds. 1985.

Marshall, A. J. 1954. *Bower-birds: Their displays and breeding cycles.* Oxford: Clarendon Press.

Mayr, E. 1935. Bernard Altum and the territory theory. *Proc. Linn. Soc. N.Y.* Nos. 45–46:24–38.

Meanley, B. 1955. A nesting study of the Little Blue Heron in eastern Arkansas. *Wilson Bull.* 67:84–99.

Mock, D. W. 1976. Pair-formation displays of the Great Blue Heron. *Wilson Bull.* 88:185–230.

Myers, J. P. 1983. Space, time, and the pattern of individual associations in a group-living species: Sanderlings have no friends. *Behav. Ecol. Sociobiol.* 12: 129–134.

Nagata, H. 1986. Female choice in Middendorff's Grasshopper-Warbler. *Auk* 103:694–700.

Newbigin, M. I. 1898. *Colour in nature: A study in biology.* London: John Murray.

Payne, R. B. 1984. Sexual selection, lek and arena behavior, and sexual size dimorphism in birds. *Amer. Ornith. Union, Ornith. Monogr.* No. 33:1–52.

Pepperberg, I. M. 1985. Social modeling theory: A possible framework for understanding avian vocal learning. *Auk* 102:854–864.

Power, D. M. 1967. Epigamic and reproductive behavior of Orange-chinned Parakeets in captivity. *Condor* 69:28–41.

Pruett-Jones, M. A., and S. G. Pruett-Jones. 1985. Food caching in the tropical frugivore, MacGregor's Bowerbird (*Amblyornis macgregoriae*). *Auk* 102:334–341.

Pruett-Jones, S. G., and M. A. Pruett-Jones. 1988a. A promiscuous mating system in the Blue Bird of Paradise *Paradisaea rudolphi. Ibis* 130:373–377.

———. 1988b. The use of court objects by Lawes' Parotia. *Condor* 90:538–545.

Prum, R. O. 1985. Observations of the White-fronted Manakin (*Pipra serena*) in Suriname. *Auk* 102:384–387.

———. 1986. The displays of the White-throated Manakin *Corapipo gutturalis* in Suriname. *Ibis* 128:91–102.

Prum, R. O., and A. E. Johnson. 1987. Display behavior, foraging ecology, and systematics of the Golden-winged Manakin (*Masius chrysopterus*). *Wilson Bull.* 99:521–539.

Pugesek, B. H., and K. L. Diem. 1986. Ages of mated pairs of California Gulls. *Wilson Bull.* 98:610–612.

Putnam, L. S. 1949. The life history of the Cedar Waxwing. *Wilson Bull.* 61:141–182.

Rand, A. L. 1938. Results of the Archbold Expeditions, No. 22: On the breeding habits of some birds of paradise in the wild. *Amer. Mus. Novitates* 993:1–8.

———. 1940. Results of the Archbold Expeditions, No. 26: Breeding habits of the birds of paradise: *Macgregoria* and *Diphyllodes. Amer. Mus. Novitates* 1073:1–14.

Rice, D. W., and K. W. Kenyon. 1962. Breeding cycles and behavior of Laysan and Black-footed albatrosses. *Auk* 79:517–567.

Richards, O. W. 1927. Sexual selection and allied problems in insects. *Biol. Rev. Cambridge Philos. Soc.* 2:298–360.

Richdale, L. E. 1951. *Sexual behavior in penguins.* Lawrence: University of Kansas Press.

Ripley, D. 1942. *Trail of the money bird.* New York: Harper and Brothers.

Robbins, M. B. 1983. The display repertoire of the Band-tailed Manakin (*Pipra fasciicauda*). *Wilson Bull.* 95:321–342.

Rodgers, J. A. Jr. 1977. Breeding displays of the Louisiana Heron. *Wilson Bull.* 89:266–285.

Rothschild, M., and C. Lane. 1960. Warning and alarm signals by birds seizing aposematic insects. *Ibis* 102:328–330.

Rowley, I. 1975. *Australian bird life.* Sydney: Collins.

Santayana, G. 1896. *The sense of beauty: Being the outline of aesthetic theory.* New York: Charles Scribner's Sons. Dover reprint, 1955.

Savard, J.-P. L. 1985. Evidence of long-term pair bonds in Barrow's Goldeneye (*Bucephala islandica*). *Auk* 102:389–391.

Schwartz, P., and D. W. Snow. 1979. Display and related behavior of the Wire-tailed Manakin. *Living Bird* 17(for 1978):51–78.

Scott, J. W. 1942. Mating behavior of the Sage Grouse. *Auk* 59:477–498.

Selous, E. 1927. *Realities of bird life.* London: Constable and Co.

Shaw, P. 1985. Age-differences within pairs of Blue-eyed Shags *Phalacrocorax atriceps. Ibis* 127:537–543.

Shepard, J. M. 1976. Factors influencing female choice in the lek mating system of the Ruff. *Living Bird* 14(for 1975):87–111.

Sick, H. 1954. Zur Biologie des amazonischen Schirmvogels, *Cephalopterus ornatus. J. Orn.* 95:233–244.

———. 1959. Die Balz der Schmuckvögel. *J. Orn.* 100:269–302.

———. 1967. Courtship behavior in the manakins (Pipridae): A review. *Living Bird* 6:5–22.

———. 1984. *Ornitologia brasileira: Uma introdução.* Brasilia: Editora Universidade de Brasilia.

Silberglied, R. E. 1984. Visual communication and sexual selection among butterflies. In Vane-Wright and Ackery 1984.

Skutch, A. F. 1949. Life history of the Yellow-thighed Manakin. *Auk* 66:1–24.

———. 1954. *Life histories of Central American birds,* vol. 1. Pacific Coast Avifauna 31. Berkeley, Calif.: Cooper Ornithological Society.

———. 1960. *Life histories of Central American birds,* vol. 2. Pacific Coast Avifauna 34. Berkeley, Calif.: Cooper Ornithological Society.

———. 1964a. Life histories of hermit hummingbirds. *Auk* 81(5):25.

———. 1964b. Life history of the Scaly-breasted Hummingbird. *Condor* 66:186–198.

———. 1967. *Life histories of Central American highland birds.* Publ. Nuttall Ornith. Club 7. Cambridge, Mass.

———. 1968. The nesting of some Venezuelan birds. *Condor* 70:66–82.

———. 1969. *Life histories of Central American birds,* vol. 3. Pacific Coast Avifauna 35. Berkeley, Calif.: Cooper Ornithological Society.

———. 1970. The display of the Yellow-billed Cotinga, *Carpodectes antoniae. Ibis* 112:115–116.

———. 1971. *A naturalist in Costa Rica.* Gainesville: University of Florida Press.

———. 1972. *Studies of tropical American birds.* Publ. Nuttall Ornith. Club 10. Cambridge, Mass.

———. 1973. *The life of the hummingbird.* New York: Crown Publishers.

———. 1976. *Parent birds and their young.* Austin: University of Texas Press.

———. 1977. *A bird watcher's adventures in tropical America.* Austin: University of Texas Press.

———. 1980a. *A naturalist on a tropical farm.* Berkeley: University of California Press.

———. 1980b. Arils as food of tropical American birds. *Condor* 82:31–42.

———. 1981. *New studies of tropical American birds.* Publ. Nuttall Ornith. Club 19. Cambridge, Mass.

———. 1983a. *Birds of tropical America.* Austin: University of Texas Press.

———. 1983b. *Nature through tropical windows.* Berkeley: University of California Press.

———. 1985. *Life ascending.* Austin: University of Texas Press.

———. 1987a. *Helpers at birds' nests: A worldwide survey of cooperative breeding and related behavior.* Iowa City: University of Iowa Press.

———. 1987b. *A naturalist amid tropical splendor.* Iowa City: University of Iowa Press.

Slud, P. 1957. The song and dance of the Long-tailed Manakin, *Chiroxiphia linearis. Auk* 74:333–339.

Smith, D. A. S. 1984. Mate selection in butterflies: Competition, coyness, choice and chauvinism. In Vane-Wright and Ackery 1984.

Snow, B. K. 1961. Notes on the behavior of three cotingas. *Auk* 78:150–161.

———. 1970. A field study of the Bearded Bellbird in Trinidad. *Ibis* 112:299–329.

———. 1972. A field study of the Calfbird, *Perissocephalus tricolor. Ibis* 114:139–162.

———. 1973a. Social organization of the Hairy Hermit *Glaucis hirsuta. Ardea* 61:94–105.

———. 1973b. The behavior and ecology of hermit hummingbirds in the Kanaku Mountains, Guyana. *Wilson Bull.* 85:163–177.

———. 1974. Lek behaviour and breeding of Guy's Hermit Hummingbird. *Ibis* 116:278–297.

———. 1977. Territorial behavior and courtship of the male Three-wattled Bellbird. *Auk* 94:623–645.

Snow, B. K., and D. W. Snow. 1985. Display and related behavior of male Pintailed Manakins. *Wilson Bull.* 97:273–282.

Snow, D. W. 1956. The dance of the manakins. *Anim. Kingdom* 59:86–91.

———. 1961. The displays of the manakins *Pipra pipra* and *Tyranneutes virescens. Ibis* 103a:110–113.

———. 1962a. A field study of the Black and White Manakin, *Manacus manacus,* in Trinidad. *Zoologica* (N.Y. Zool. Soc.) 47(2):65–104.

———. 1962b. A field study of the Golden-headed Manakin, *Pipra erythrocephala,* in Trinidad. *Zoologica* (N.Y. Zool. Soc.) 47(4):183–198.

———. 1963a. The display of the Orange-headed Manakin. *Condor* 65:44–48.

———. 1963b. The evolution of manakin displays. *Proc. 13th Internatl. Ornith. Congr.,* pp. 553–561.

———. 1963c. The display of the Blue-backed Manakin, *Chiroxiphia pareola,* in Tobago, W.I. *Zoologica* (N.Y. Zool. Soc.) 48(4):167–176.

———. 1968. The singing assemblies of Little Hermits. *Living Bird* 7:47–55.

———. 1971. Social organization of the Blue-backed Manakin. *Wilson Bull.* 83:35–38.

————. 1976. *The web of adaptation.* New York: Quadrangle, N.Y. Times Book Co.

————. 1977. The display of the Scarlet-horned Manakin, *Pipra Cornuta. Bull. British Ornith. Soc.* 97:23–27.

————. 1982. *The cotingas.* Tring: British Museum (Natural History); Oxford: Oxford University Press.

Snow, D. W., and B. K. Snow. 1973. The breeding of the Hairy Hermit *Glaucis hirsuta* in Trinidad. *Ardea* 61:106–122.

Stewart, P. A. 1959. The "romance" of the Wood Duck. *Audubon Mag.* 61: 63–65.

Stiles, F. G. 1973. Food supply and annual cycle of the Anna Hummingbird. *Univ. California Publ. Zool.* 97:1–109.

————. 1982. Aggressive and courtship displays of the male Anna's Hummingbird. *Condor* 84:208–225.

Stiles, F. G., and L. L. Wolf. 1979. Ecology and evolution of lek mating behavior in the Long-tailed Hermit Hummingbird. *Amer. Ornith. Union, Ornith. Monogr.* No. 27:1–78.

Stonor, C. R. 1940. *Courtship and display among birds.* London: Country Life.

Storer, R. W. 1963. Courtship and mating behavior and the phylogeny of grebes. *Proc. 13th Internatl. Ornith. Congr.,* pp. 562–569.

————. 1967. Observations on Rolland's Grebe. *Hornero* 10:339–350.

————. 1969. The behavior of the Horned Grebe in spring. *Condor* 71:180–205.

————. 1982. The Hooded Grebe on Laguna de los Escarchados: Ecology and behavior. *Living Bird* 19:51–67.

Thorpe, W. H. 1958. The learning of song patterns by birds, with special reference to the songs of the Chaffinch *Fringilla coelebs. Ibis* 100:535–570.

Tinbergen, N. 1951. *The study of instinct.* Oxford: Clarendon Press.

————. 1958. *Curious naturalists.* London: Country Life.

Trail, P. W. 1987. Predation and antipredator behavior at Guianan Cock-of-the-Rock leks. *Auk* 104:496–507.

Urquhart, F. A. 1960. *The Monarch butterfly.* Toronto: University of Toronto Press.

Vane-Wright, R. I., and P. R. Ackery. 1984. *The biology of butterflies.* London and New York: Academic Press.

Van Someren, V. G. L. 1956. *Days with birds: Studies of some East African species.* Fieldiana: Zoology 38. Chicago: Chicago Natural History Museum.

Vaughan, R. 1980. *Plovers.* Lavenham, Suffolk: Terence Dalton.

Vellenga, R. E. 1970. Behaviour of the male Satin Bower-bird at the bower. *Australian Bird-Bander,* March:3–11.

————. 1980. Distribution of bowers of the Satin Bowerbird at Leura, NSW., with notes on parental care, development, and independence of the young. *Emu* 80:97–102.

Verner, J. 1976. Complex song repertoire of Long-billed Marsh Wrens. *Living Bird* 14(for 1975):263–300.

Von Frisch, K. 1954. *The dancing bees: An account of the life and senses of the honey bee.* London: Methuen and Co.

Wagner, H. O. 1946. Observaciones sobre la vida de *Calothorax lucifer*. *Anales Inst. Biol.* (Mexico) 17:283–299.

———. 1954. Versuch einer Analyse der Kolibribalz. *Zeit. Tierpsychologie* 11: 182–212.

Wallace, A. R. 1871. *On natural selection*. 2d ed. New York: Macmillan.

———. 1872. *The Malay Archipelago*. 4th ed. London: Macmillan and Co.

Warham, J. 1957. Notes on the display and behaviour of the Great Bower-bird. *Emu* 57:73–78.

———. 1958. The nesting of the Pink-eared Duck. *Wildlife Trust Ninth Annual Report*. Pages not numbered.

———. 1962. Field notes on Australian bower-birds and cat-birds. *Emu* 62: 1–30.

———. 1963. The Rockhopper Penguin, *Eudyptes chrysocome*, at Macquarie Island. *Auk* 80:229–256.

Weed, C. M. 1923. *Butterflies worth knowing*. Garden City, N.Y.: Doubleday, Page and Co.

Weller, M. W. 1968. Notes on some Argentine Anatidae. *Wilson Bull.* 80:189–212.

Wickler, W. 1968. *El mimetismo en las plantas y en los animales*. Mexico City: McGraw-Hill Book Co.

Wiese, J. H. 1976. Courtship and pair formation in the Great Egret. *Auk* 93: 709–724.

Wiley, R. H. 1971. Song groups in a singing assembly of Little Hermits. *Condor* 73:28–35.

———. 1973. Territoriality and non-random mating in Sage Grouse *Centrocercus urophasianus*. *Anim. Behav. Monogr.* 6:85–169.

Williams, D. M. 1983. Mate choice in the Mallard. In Bateson, ed. 1983.

Willis, E. O. 1966. Notes on a display and nest of the Club-winged Manakin. *Auk* 83:475–476.

———. 1967. The behavior of Bicolored Antbirds. *Univ. California Publ. Zool.* 79:1–127.

Index

Illustrations are indicated by boldfaced page numbers.